The Stakeholder Perspective

The Stakeholder Perspective

Relationship Management to Increase Value and Success Rates of Projects

By Massimo Pirozzi

CRC Press is an imprint of the
Taylor & Francis Group, an **informa** business

CRC Press
Taylor & Francis Group
6000 Broken Sound Parkway NW, Suite 300
Boca Raton, FL 33487-2742

Printed on acid-free paper

International Standard Book Number-13: 978-0-367-18476-6 (Hardback)

Visit the Taylor & Francis Web site at
http://www.taylorandfrancis.com

and the CRC Press Web site at
http://www.crcpress.com

Contents

PART II THE RELATIONSHIP MANAGEMENT PROJECT

Foreword

Stakeholders, those who influence or care about something, is such a simple concept. Yet it has taken nearly 50 years for those involved with planning and managing projects to fully understand the importance of stakeholders to project management. In recent years, stakeholder management has been added to project management standards, guides, and best practices. The topic has grown to include a wide range of stakeholder engagement-related activities and processes; it is the subject of numerous books, articles, reports, and studies. Stakeholder engagement, or stakeholder management, is included in most project management plans, systems, and approaches today. However, I don't think its importance is yet fully appreciated.

Project stakeholders include the individuals and organizations that plan, finance, implement, oversee, and complete every project. They include shareholders, customers, users, and anyone who benefits from a project. And they often include those affected by a project, including neighbors, government regulators, and many citizens. In my opinion, stakeholder attitudes, expectations, and perceptions may be the most important aspect of managing any project, more important than scope, schedule, cost, or anything else. Of course, all the standard elements of project management are important, but without the satisfaction of key stakeholders, a project cannot be successful.

Perhaps the best-known example of this was the project to build the Sydney Opera House in Australia. The project far exceeded original scope, cost and time estimates; it was a failure by nearly all established measures. Yet it became one of the most iconic and popular buildings in the world. The completed project was loved by its most important stakeholders—the people who saw and visited it. Not only was it a unique and beautiful building, but it also helped attract millions of visitors to Sydney who have contributed hundreds of millions of dollars to the local economy. It was a fantastic success for one main reason—stakeholders!

Like many other simple concepts, however, effective project stakeholder engagement is not so easy. Organizing and working with other human beings can always be a challenge. Dealing with many different groups of people can be complicated and difficult. In fact, large and diverse sets of stakeholders are now widely recognized as a major element of project complexity. It is sometimes difficult to even identify all project stakeholders, let alone understand their attitudes, expectations, and perceptions. In fact, some stakeholders might not know how they feel about a project, which can be both a challenge and an opportunity for project managers. Some may not even know they are project stakeholders; when that awareness comes, however, some can become very important stakeholders indeed—for example, those who may be negatively affected by the project or the product of the project. Project managers (and project sponsors) ignore those stakeholders at their own peril.

Stakeholder engagement is a critical success factor for every project. Stakeholder identification, analysis, engagement, communications, and relationship management should be a major aspect of managing every project. Of course, these issues also apply to program and project portfolio management. It is not necessarily easy, but it is not mysterious either. It is just very important!

This book is an evidence about Massimo Pirozzi's understanding of the importance of stakeholders to project success. It also goes beyond other books on this topic in several significant areas. Stakeholder networks are introduced, the use of interpersonal skills in interacting with stakeholders is explained, and the important topics of ethics and value creation in stakeholder relations are emphasized. The relationship of stakeholders and project complexity is discussed at length and a process for successful stakeholder engagement to overcome that complexity is provided. In our rapidly changing and increasingly complex world, the probability of project success can be dramatically increased with the use of the concepts contained in this book.

Massimo has demonstrated his commitment to stakeholders throughout his career, and in his professional leadership and recent writings. Now he is sharing more of his understanding and knowledge with the rest of us. I am honored to author this foreword. I feel even more honored due to the importance of this topic. Please read this book. It's great! Then put it to good use.

David L. Pells
Editor/Publisher
PM World Journal, PM World Library
PMI Fellow, Honorary Fellow of ISIPM (Italy),
APM (UK), PMA (India), SOVNET (Russia)
Addison, Texas, USA

Preface

This book focuses on the centrality of people—the stakeholders, in both projects and project management. In short, this centrality is because stakeholders are both the doers, and the beneficiaries, of each project, and, therefore, they contribute to projects' success in all respects: moreover, stakeholders are, at the same time, the greatest generators both of the value to be delivered, and of the complexity to be faced and solved. Since, despite the recent improvements due to the increasing maturity in project management of the organizations—which are, in any case, partially counterbalanced by the continuously increasing complexity of projects, there is still an actual situation, in which important percentages of projects do not meet their original goals/business intents, i.e., they do not satisfy their stakeholders' expectations, and/or, moreover, they experience scope creeps, cost overruns, and time delays, i.e., they do not correspond to initial project/stakeholder requirements too, a further attention to stakeholders and to stakeholder relationships domain is absolutely appropriate in order to improve performances, especially in large and/or complex projects.

The purpose of this book is, therefore, to propose to the project management community, a helpful, innovative, stakeholder-centered approach, to increase both the delivered value and the success rate in projects of all size and complexity: this approach is declined in a logical model, which is called the "Stakeholder Perspective", and then, it is structured, in order to be more effective, in an immediately manageable "project in the project", which is called "the Relationship Management Project". Ultimately, an additional purpose of this book is to provide a consistent, complete but synthetic, updated focus on both stakeholder concepts and stakeholder identification/analysis/management processes. Above focus, which can also be considered as a deepening and/or an integration of the existing literature about stakeholder management, on the one hand, includes several innovative issues but, on the other hand, maintains a strict coherence with project management international standards and best practices.

The innovative stakeholder-centered journey toward the increase of the projects' effectiveness, efficiency, and success rate, which is proposed in this book, is structured in two parts. Part I looks into the stakeholder perspective in both projects and project management, starting with the investigation about the nature and the role of the stakeholders, then deepening, and integrating with some significant innovations, the domains of project management processes that focus on stakeholders, including stakeholder identification, stakeholder analysis, management of the relationships with key stakeholders, and stakeholder network management, to conclude with an overview about those personal and interpersonal skills, and ethics, which are basic in stakeholder relationships. Part II explores several innovative issues, starting with the indissoluble link between stakeholder relations and delivered value, and with the satisfaction of both stakeholder requirements and expectations as the critical success factor in all projects, then proceeding with facing successfully different levels of project complexity, and with targeting both project and business value generation by using KPIs, in order to definitively propose an innovative structured path to effectiveness called Relationship Management Project. Finally, in the last chapter, some innovative cues about Project Management X.0, a possible stakeholder-centered evolution for both project and portfolio management, are provided for the reader's attention.

The above contents are proposed in the light format of a guide, with the purpose of being more easily readable and usable: furthermore, in order to enhance immediate applicability, almost all contents that are contained in this guide, including innovative issues and models, should be considered as actual, because they are the results of direct experiences and/or observations.

Finally, since we are all stakeholders, this book is dedicated to all of us.

Acknowledgments

Many thanks to Russ D. Archibald, a globally recognized author, consultant and lecturer on project management, and to Alan Stretton, one of the pioneers of modern project management, for their several positive commentaries about my previous research works, which encouraged me greatly to proceed with this book.

Many thanks also to my friend David L. Pells, real project management globalist, especially for having been my first international editor, who immediately believed in my research work.

Several thanks to the members of my Project Management Association, the Istituto Italiano di Project Management (ISIPM), which first gave me the opportunity, about 5 years ago, to present in a 3-hour seminar the results of my first researches on stakeholders: special thanks to my friends and fellow travelers in project management, Enrico Mastrofini, Graziano Trasarti, Vito Introna, Maurizio Monassi, and Biagio Tramontana, mostly for all the great job made together, to Federico Minnelle, for having always been a real scientific reference, and to my friend Alessandro Quagliarini, particularly for his encouragements to become a teacher, which today is my principal work.

Thanks to all over thousand people I directly dealt with in my professional life, both as a manager and as a teacher, for having made real all the experiences, both positive and negative, which are the basis of this book.

The greatest thanks to my awesome wife Antonella, my prime supporter, and to my fantastic sons Paolo and Marco, especially for their continuous closeness, which has been foundational for me to write this book.

About the Author

Massimo Pirozzi, MSc (cum laude), Electronic Engineering, University of Rome "La Sapienza", principal consultant, project manager, lecturer and educator. He is a member and the secretary of the Executive Board, a member of the Scientific Committee, and an accredited master teacher of the Istituto Italiano di Project Management (Italian Institute of Project Management). He is certified as a professional project manager, as an information security management systems lead auditor, and as an international mediator. He is a researcher and an author about stakeholder relationship management, complex projects management, and effective communication. Massimo has a wide experience in managing large and complex projects in national and international contexts, and in managing relations with public and private organizations, including multinational companies, small and medium-sized enterprises, research institutes, and non-profit organizations. He worked successfully in several sectors, including defense, security, health, education, cultural heritage, transport, gaming, and services to citizens. He was also, for many years, a top manager in ICT industry, and an adjunct professor in organizational psychology. He is registered as an expert of the European Commission, and as an expert of the Italian Public Administrations. Massimo is the international correspondent for the PM World Journal and the PM World Library in Italy.

THE STAKEHOLDER PERSPECTIVE

Chapter 1

Stakeholders, Who Are They?

Stakeholders are *persons*, without a doubt: but why do we use a word that specific, which incorporates so many concepts that some hundreds of its different definitions exist in the literature, and direct translation of which in other languages is nearly impossible? In fact, some history, and a little in-depth analysis, of the meanings of this word can help us to reveal a significant part of the mystery.

The word *stakeholder* dates back to the beginning of the eighteenth century, in England, and it meant the person who was entrusted with the *stakes* of bettors: he was the *holder* of all the bets placed on a game or a race, and he was the one who was paying the money to the winners. Therefore, *the first stakeholder was a holder of interests*, and this is, even today, one of the most common meanings, if we consider, in addition, that "having a stake" is a synonym of "having an interest", and that *stakes* (meaning "strong sticks") can be pushed in the ground either to mark a property, or to be part of a fence that settles the boundaries of an estate, so defining the perimeter of an interest.

But stakes (still meaning "strong sticks") can be hammered in the ground also for *supporting* plants: in fact, it is believed that the first modern meaning of stakeholders, which has been attributed (Freeman, 1984) to an internal memorandum of Stanford University Research Center dated 1963, was "*those groups without whose support the organization would cease to exist*". Therefore, stakeholders are the strongest supporters of an organization (we could also say that they may be ready to "go to the stake" for it!), and their contribution is foundational to the existence of the organization itself: main emphasis is, in this definition, on *internal* stakeholders, who are the doers of organization's performances, and who need to be properly *engaged* to give an effective contribution. Meanwhile, in the above-mentioned perspective, stakeholders do not act anymore as *individuals* only, but they are considered as *part of groups*, too: stakeholders interface each other through *processes*, start to act *collectively*, by sharing their resources and by integrating their efforts, and they do it through *relations*, so that their *behavior* becomes *organizational*, and not only *personal*.

Furthermore, in the first text on the theory of stakeholders (Freeman, 1984), the definition of stakeholder was "a stakeholder in an organization is any group or individual who can affect or is affected by the achievement of the organization's objectives". Since "to affect" is a synonym of "to influence", it was here that one other of the most common concepts in stakeholder definitions came in: stakeholders *influence* the organization's objectives, and are influenced by them, and this is the first time that the nature of stakeholders' *centrality* in the organizations became evident, since stakeholders were defined as both the actors and the recipients of the organization's results.

In the original PMBOK (Project Management Institute, 1987), the stakeholders were considered as the *participants* to the project. In addition, some years later (Freeman, 1994), the foundational concepts of *participation* and of *created value* were enhanced too, and stakeholders were defined as "*participants in the human process of joint value creation*". At first, in fact, stakeholders *participate* in the organizations, as far as other stakeholders *would like to participate* in the same organizations, and, in both cases, they want to do so *jointly*, e.g., *by becoming, and then being part of a specific community, which is deeply characterized by common goals, behavioral rules, specific languages, and so on.* Thus, regarding the created value, the extraordinary importance of this concept, on the one hand, is due to the enhancement of the stakeholders' role on the joint creation of the business and/or social value, while, on the other hand, is because it introduces the issue that *actual value which is added by stakeholders must be considered arithmetically: in fact, it can be positive, but also null or negative.* Indeed, in, and outside of, all the organizations, there are *positive stakeholders*, who have to be engaged, and whose needs have to be satisfied, but there are generally also *neutral/reluctant stakeholders*, who require special additional efforts for their engagement, and *negative/hostile stakeholders* too, who have to be *disengaged*, if possible, but, in any case, must be *dissatisfied*, in order to achieve organization's goals.

Moreover, starting from the second half of the 1980s, *a vision of the enterprise as a complex system inserted in the society was born*, and a new "*stakeholder theory*" started to be developed. In this vision, the basic idea is to extend the benefits that are created by the enterprise from its shareholders to its participants to various titles, and to affirm the conviction that this same idea must be translated in ethical principles of managerial behavior. Without entering into the dispute between stakeholder view and shareholder view (also because shareholders are stakeholders, too), it is important to notice that above focus on *corporate social responsibility* started to incorporate an important *ethical* component into the concept of stakeholder. In fact, it is confirmed that a strong ethical concept is still valid today, since, for instance, Cambridge Dictionary defines stakeholder as "a person such as an employee, customer, or citizen who is involved with an organization, society, etc., and therefore has *responsibilities* toward it and an interest in its success": conveniently, *responsibility* is considered by the community of project managers *as one of the top ethical issues* (Project Management Institute, 2006).

Finally, if we go back to the word *stake*, it could be interesting to notice that one of its main meaning is *risk*, and that "at stake" is a synonym of "at risk": in other words, *stakeholders do risk in order to achieve their goals*, and *risk-based thinking* is part of their life, exactly as it is part of several disciplines, from project management, since its early beginning, to quality management, more recently, and so on. On the other side, *stakeholders are the ones who introduce risk in all the domains.*

Are all the above considerations applicable to *project stakeholders*? Of course they are, with a specific focus on *unicity*, and a necessary emphasis on the relation between *stakeholder expectations* and *project goals*, which confirms the *stakeholder centrality in all projects.* Every project is, actually, unique, and its *unicity* is reflected not only in its scope, goals, objectives, deliverables, time, cost, resources, and so on, but also in its own *set of stakeholders*, which, then, characterizes specifically each project both with respect to others—also if they may be related in the same program, and in terms of its inherent complexity. Furthermore, a foundational issue is that *stakeholders are central in all the projects.* In fact, organizations define strategies, which are based on their own mission and vision, and *projects are means to accomplish strategic goals*, then achieving, through their results, the expected benefits: *the overall value that is generated by each project determines the stakeholder satisfaction, and the relevant project success rate, within the whole investment life cycle* (see Figure 1.1). It is, indeed, a fact, that each project exists to implement an *investment*, which, in turn, has been mutually agreed to *harmonize different stakeholder expectations*: organizations define strategies, which are based on their own mission and vision, then select pursuable opportunities in accordance with their defined strategy, then set business cases up, and, finally, start projects up.

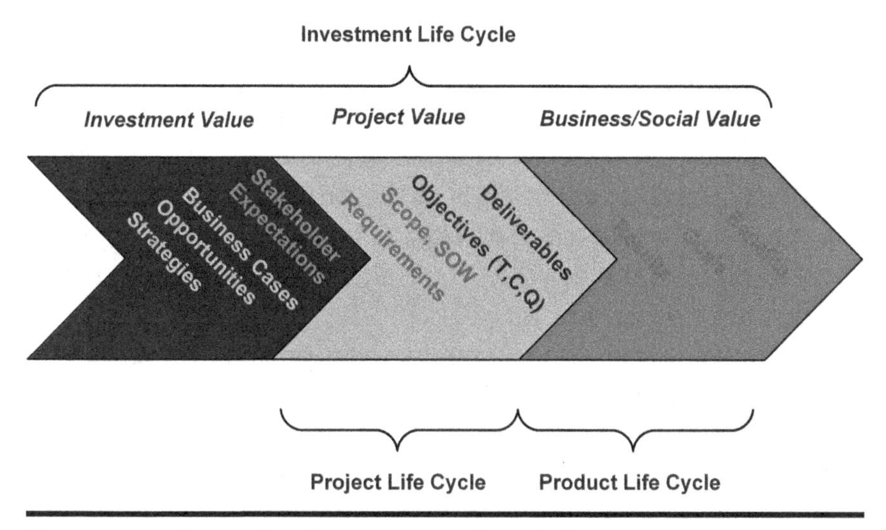

Figure 1.1 The project investment value chain.

The inputs of a project generally include, then, business case, contract, and Statement of Work, or things like that: of course, there are different business cases or similar for different stakeholders, as, for instance, providers, investors, and customers are, and this leads to the *existence of different perspectives, in terms of results to be achieved, that will accompany the project in all its life cycle*, and also afterwards, i.e., in released product/infrastructure/service life cycle.

In fact, while, on the one side, business cases, which are the causes of project start-up, are based on *stakeholder business expectations*, which, in turn, correspond to *project goals to be achieved*, on the other side, contract and SOW (Statement Of Work), which are the references for project development and delivery, are based on *stakeholder requirements, which are, in turn, the conversion of different stakeholder expectations in a commonly agreed (at least initially) project scope, and which correspond to project objectives to be delivered.* Project manager, project team, and other stakeholders implement the project in accordance with the requirements, and, then, deliver it to other stakeholders: *project can be considered really successful when its goals are realized*, then *achieving those results that correspond to the stakeholder expectations in terms of created business value and relevant achieved benefits.*

This double role of stakeholders is the pillar of all projects, and confirms the stakeholder centrality: stakeholders are both the doers and the beneficiaries of the project, and while, on the one side, the stakeholder work is basic for project implementation, on the other side, the stakeholder satisfaction is the main project success factor.

Definitively, a project stakeholder is a person, or a group of persons, or an organization, who

- *participates*, or *would like to participate*, in the project;
- has some kind of *interest* in the project;
- can be (if properly *engaged*) a foundational *supporter* of the project;
- may *affect/influence* the project, or may *be affected/ influenced* by the project itself;
- can bring a *value*, which could be either positive or negative, to the project;
- may have *responsibilities* toward the project, which, in turn, is supposed to *satisfy his requirements and expectations*;
- is characterized by a *risk*-based thinking approach;
- is part of a set that characterizes *uniquely* each project; and
- has a *central role in all projects*: stakeholders, indeed, both implement the project and determine its success via their satisfaction, and, then, are the actual key for project success.

It is, then, a fact, on the one hand, that *each project includes a large variety of stakeholders*, having different interests, expectations, level of influence, responsibilities, etc., and, on the other hand, that, due to the key centrality of their role, *all stakeholders are important.*

In project management literature, the definitions of project stakeholders are quite essential: while in first PMBOK (Project

Management Institute, 1987), stakeholders were considered as just the synonym of "participants", maybe the first real definition of project stakeholders, with an evident emphasis on *active involvement* and *interest* aspects, was "individuals and organizations who are actively involved in the project, or whose interests may be positively or negatively affected as a result of project execution or successful project completion" (Project Management Institute, 1996). However, current definitions leave the active involvement issue, which, on one part, limits the stakeholder domain, while, on the other, is more an objective to be reached via positive *engagement*, rather than an assumption that is part of a definition, and, at the same time, insist on *influence* aspects as the core ones. Moreover, some perspectives add the essential human factor of *perception* "stakeholder is a person, group or organization that has interests in, or can affect, be affected by, or *perceive* itself to be affected by, any aspect of the project" (International Organization for Standardization, 2012), some other insist on the first concept of *participation* "all individuals, groups or organisations *participating* in, affecting, being affected by, or interested in the execution or the result of the project can be seen as stakeholders" or, very similarly, relaxing the *interest* issue but with a further extension to programs and portfolios concepts, "stakeholder is an individual, group or organization that may affect, be affected by, or perceive itself to be affected by a decision, activity, or outcome of a project, program, or portfolio" (Project Management Institute, 2017).

In general, typical project stakeholders include two groups that are both temporary and specific structures of the project domain, i.e., the project organization and the project governance stakeholders, and other groups, as investors, customers, and special interests groups, that we can find also in more stable structures, as organizations generally are. The project organization is the temporary structure that includes (International Organization for Standardization, 2012) the project manager and the project team, and that may include a

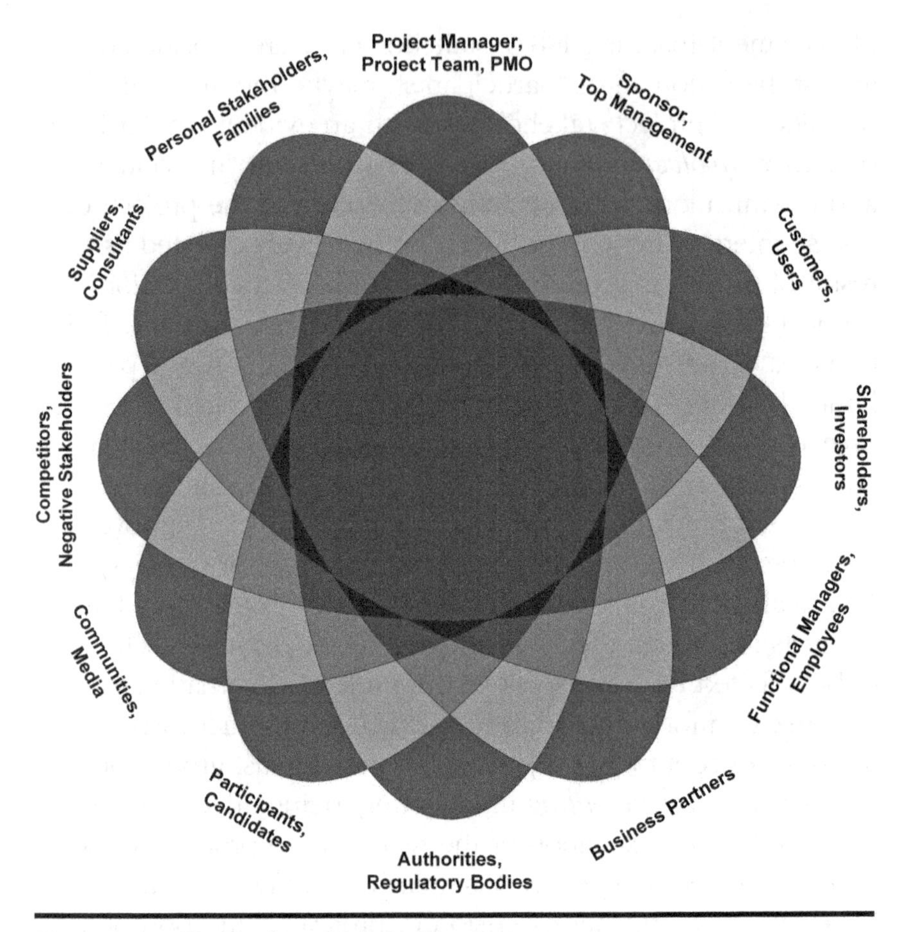

Figure 1.2 The Project Stakeholder Rose.

project management team and a Project Management Office, too, while project governance may involve the project sponsor, a project steering committee or board, and other top management executives. Definitively, since all the stakeholders are important, it is foundational to consider them all properly.

An accurate overview about different project stakeholders should include (see Figure 1.2):

- the project manager;
- the project team, the project management team, the project management office;

- the sponsor, the project steering committee or board, the top management;
- the customers, the users, the contracting officers;
- the shareholders, the investors, the funders, the partners;
- the functional and/or resource managers, the employees, the professionals, the collaborators;
- the business partners, the network partners, the distributors, the representatives, the members of the consortium;
- the suppliers, the consultants, the service companies, the outsourcers;
- the authorities, the central and local public administration, the regulatory bodies;
- the potential customers and users, the participants and the candidates to participate in the project;
- the local communities, the web communities, the associations, the trade unions, the media;
- the competitors, and the other reluctant and/or negative and/or hostile stakeholders; and
- the personal stakeholders, including stakeholders' families, lovers, close friends, and generally, all the persons the stakeholders have strong personal relations with.

Project managers, who are they? In the project management literature, one of the first definitions of project manager was "*the individual responsible for managing the project*" (Project Management Institute, 1996), and it is interesting to notice that, in any case, *responsibility* has been recognized by the community of project managers as the most important *ethical value* to be used as a primary reference (Project Management Institute, 2006). In today's definitions, there is a special focus on *leadership*, too: in fact, project manager either "*leads* and manages project activities and is accountable for project completion" (International Organization for Standardization, 2012), or "is the person assigned by the performing organization to *lead* the team that is responsible for achieving the project objectives" (Project Management Institute, 2017). However,

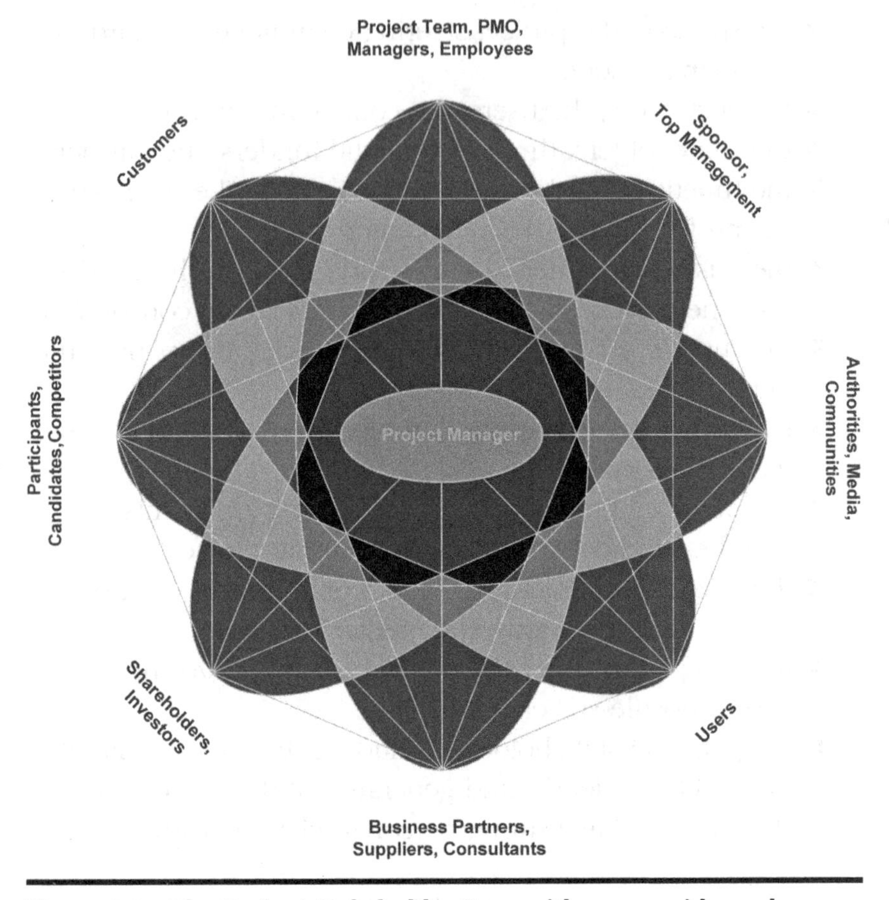

Figure 1.3 The Project Stakeholder Rose with some evidenced relations among stakeholder groups.

furthermore, in modern project management, project manager's role(s) and competences became, and are still becoming, so important, that *project manager can be considered, in his own full right, the central stakeholder among stakeholders.*

At first, in fact, *project manager is the only stakeholder who must have relations with all other project stakeholders* (see Figure 1.3), and, as we will see in following chapters, *who is responsible of monitoring relations among other stakeholders, too.* The immediate, although quite never enough emphasized, consequences of this project manager's role are that, on the one hand, *communication with stakeholders absorbs the largest part of project manager's time* (literature reports percentages that are higher than 80%, maybe about 90%, and that are still increasing), while, on the other hand, in *the major part of project management processes, the management of relations with stakeholders is needed*, as we will demonstrate in the next chapter.

Moreover, *each modern project manager must be able to fulfill different high-level roles*, which, on the one hand, reflect his capabilities, and, on the other hand, are representative of the value and contributions of his profession, within his sphere of influence; specifically, these roles include (Project Management Institute, 2017):

- in project's domain, the project manager has to lead the project team to meet the project's objectives and stakeholders' expectations, but also he has to work to balance the competing constraints on the project with the resources available, and he must perform a crucial communication role between the project sponsor, team members, and other stakeholders, too;
- in the organization's domain, the project manager proactively interacts with other project managers, and maintains a strong advocacy role within the organization;
- in the industry's domain, the project manager stays informed about current industry trends, takes this information, and sees how it may impact or apply to the current projects;
- in the professional discipline's domain, the project manager has to take care of continuous knowledge transfer and integration; and
- in domains across disciplines, the project manager can have a basic role of educator, and a professional project manager may choose to orient and educate other professionals regarding the value of a project management approach to the organization.

Definitely, project manager is the unique stakeholder who spends almost 90% of his time by acting as a front office versus the other stakeholders (while only about 10% or less of his time is absorbed by back office activities, mainly in planning and controlling processes), and, then, he actually can be considered as the stakeholder who is central with respect to all other stakeholders.

It can be interesting to notice that *above time percentages*, which are valid for project managers, *are almost inverted in the case of project team members*, who, on the contrary, dedicate the major part of their time to perform "technical" project activities, and, then, have relatively little time to dedicate to relations with other stakeholders: although this time is almost spent internally, i.e., with the project manager and with other team members, it is foundational externally too, in order both to check the validity of the requirements with customers and/or users, and to valorize the perceived quality.

Finally, while existing literature substantially agrees on the belonging to the project stakeholders' domain of almost all individuals and groups that have been identified as those that are relevant for the project, including project manager, the project team, the project management team, the project management office, the sponsor, the project steering committee or board, the top management, the customers, the users, the contracting officers, the shareholders, the investors, the funders, the partners, the functional and/or resource managers, the employees, the professionals, the collaborators, the business partners, the network partners, the distributors, the representatives, the members of the consortia, the suppliers, the consultants, the service companies, the outsourcers, the authorities, the central and local public administration, the regulatory bodies, the potential customers and users, the participants and the candidates to participate in the project, the local communities, the web communities, the associations, the trade unions, the media, the web, *there are two groups of foundational importance that have been, quite incomprehensibly, substantially forgotten and/or neglected by project management literature: the reluctant/negative/hostile stakeholders, and the personal stakeholders, i.e., the people with whom project stakeholders have strong personal relations*, including families, lovers, close friends and relatives, and so on.

Reluctant stakeholders are not necessarily hostile, but, since they tend not to give their expected support, they do not guarantee their contribution in terms of positive value to the project: special additional efforts are then continuously needed to engage them, in order to avoid possible, and frequent, negative impacts on the project itself. On the other hand, *negative and hostile stakeholders aim at bringing an increasing negative value, and, if they are not combated properly, they can become a critical issue for the project.* Both hostile and negative stakeholders have interests that are contrary to the project, and, then, hope project's failure, but, generally, *while the hostile stakeholders, as the competitors and other communities that do not want the project to be realized (e.g., the NIMBY "Not In My Back Yard" associations), are evident, the negative stakeholders, who are commonly present in the same organizations that are entitled to implement the project, and who oppose the project for reasons of internal competition/envy and/or disagreement on priorities/budgets, are generally hidden*, and their identification is often quite difficult. Furthermore, especially in Information Security domain, there are two additional categories of hidden stakeholders that are of extraordinary importance, which are the *unrevealed stakeholders* and *the two-timing stakeholders: both act as normal stakeholders until such time as they can, suddenly and/or unexpectedly, turn out to be hostile, and then create great damages to the project* through hacking, cracking, DoS (denial-of-service) attacks, social engineering, etc. Ultimately, both hostile and negative stakeholders mainly combat the project by trying to discredit it, often by using the power of amplification that is a characteristic of internet and/or social media, and, if project reputation is not properly defended, they can succeed to be disruptive for the project, as we will see in following paragraphs.

Finally, the *personal stakeholders*, who are the people with whom project stakeholders have strong personal relations, including families, lovers, close friends, and (although not always!) parents and relatives, although they are not directly involved in the project, *have an extraordinary importance in order to achieve project success.* In fact, personal stakeholders do not only wish and/or want to participate to the project, but they can be also considered as *engaged full time*, and they *can influence very significantly the motivation of key stakeholders*, including the project manager, the project team, the sponsor, the top management, the board, and the customers too. *Personal stakeholders may turn out to be either part of the best supporters of the project,* e.g., in case they believe the project is an opportunity for the key stakeholder they have a close relation with, or *part of the worst hostile/negative of the project itself,* e.g., in case *they become haters of the project because they believe that it takes up too much time to their relationship*, and, furthermore and unfortunately, quite often, on an event-based logic, *they can change rapidly, and several times, from one behavior to the other.* In project management literature, just team members' families, among all possible personal stakeholders, were mentioned as project stakeholders, but this happened only in the early beginning (Project Management Institute, 1996), while, after that, they disappeared.

Definitively, *stakeholders, i.e., people, are central with respect to all projects.* Indeed, stakeholders, including the project manager and the project team, are the *doers* of the project, as well as other stakeholders, including customers/users, and shareholders/funders, are the *target groups* of the project itself: *business is the domain in which various stakeholders interact, through both project and project management processes, to create and to exchange value. The relationships among the project stakeholders are, then, real and proper business and social relationships, which are associated with the generation, and the exchange, of business and/or social value*: in general, this flow of value, among the stakeholders, courses through the project with a continuous exchange of resources and results (Pirozzi, 2017). *Managing contents and relations of, and among, project stakeholders is the extended meaning of what we call stakeholder management*: ultimately, although only quite recently, project management community started to recognize the primary importance of both stakeholders and stakeholder management.

Chapter 2

The Recent Central Role of Stakeholders in Project Management

It is, then, a fact, that stakeholders are central with respect to all projects; however, have they been, and are they, considered central by project management discipline, too? A synthetic multiple answer could be the following:

- *stakeholders were present in project management literature from the early beginning, but they have not been considered central at all for almost 25 years.*
- *ultimately, because for very few years only, stakeholders and stakeholder management have started to be considered central.*
- *it seems that there is still a long way to pay the necessary attention to their needs and expectations, in order to satisfy them both.*

In order to explain the above multiple answers properly, some detailed analysis of main reference texts for project management, different releases of which may represent actual milestones of the *stakeholder role* that was, and is, *perceived by Project Management Community*, is needed and appropriate.

Just to start, it is interesting to notice that, while in the original PMBOK (Project Management Institute, 1987) the stakeholders were considered only as the *participants* to the project, may be the first "official" main definition of Project Management, which was contained in first edition of *PMBOK Guide* (Project Management Institute, 1996), was absolutely *stakeholder-centered*, and, at the same time, incredibly modern, since it mentioned stakeholder *expectations* too: "*Project management is the application of knowledge, skills, tools, and techniques to project activities in order to meet or exceed stakeholder needs and expectations from a project*". Immediately afterwards, *needs* were defined as *identified requirements*, while expectations were defined as *unidentified requirements*; of course, today, since there is still a trend of about the 50% of scope creeps in the projects (Project Management Institute, 2018), the purpose of "*exceeding* stakeholder needs" seems to us to be a bit anachronistic. Nevertheless, on the other side, in the same first Guide, not only *stakeholder was not included in the knowledge areas*, but, moreover, stakeholders were mentioned just in two *communication (or information?)* processes; in fact, although they were considered as a basic part of the project context, as it is today, in the whole set of processes they were mentioned only in the planning process "*communication planning*", and in the execution process "*information distribution*". In practice, among all project management processes, stakeholders started to be considered, but just as *passive* subjects, i.e., as receivers of a unilateral information distribution only; in addition, stakeholder identification was not considered as a basic process in the Initiating process group, but only as a difficult activity, and the expression *"stakeholder management" was missing at all in the Guide.* Definitively, in this first Guide, although stakeholders apparently were essential, there were no indications about managing the relations with them, except two processes of "communication (but meaning information) management"; in fact, the matter that was considered, was *not managing a complex interpersonal "communication" among stakeholders, but just a unidirectional stream of project information.*

Furthermore, four years later, in the second edition of *PMBOK Guide* (Project Management Institute, 2000), stakeholder role reached another low point, since *stakeholders disappeared from the project management main definition: "Project management is the application of knowledge, skills, tools, and techniques to project activities to meet project requirements"*. Above definition, which survived still valid and unmodified until today, i.e., almost 20 years later (Project Management Institute, 2017), *focused on requirements only*, and that *enshrined the gap between satisfying stakeholder requirements and expectations.* Second, *"project requirements", replaced "stakeholder requirements"*; although the difference at first glance seems minimal, it is not at all, because while *project requirements are evidently supposed to be objective and neutral, stakeholder requirements are intrinsically subjective and biased.* Focusing on the objectivity of project requirements, then, resolved in an *indirect suggestion both of not taking into account properly the importance of stakeholder relations in realizing deliverables, and of achieving stakeholder satisfaction in order to make the project successful*; moreover, in some way, it *incorporated also the false myth that projects have a linear behavior*, i.e., that projects, and specifically *complex* projects, can be managed purely *objectively, without taking properly into account the influence of stakeholder (people) subjectivity.* Of course, project managers know that real world is different; quite often, and especially in all complex projects, *stakeholder expectations, with their subjectivity, correct the linearity of objective project requirements*, which, for their part, *are the result of an as well as possible mediation among different stakeholder expectations* … but we will look in depth into these issues in the following chapters.

For the rest, in this edition of the *PMBOK Guide*, in terms of processes, stakeholder identification is mentioned, just once, as a difficult activity, and not as a process (*initiation* process was considered, at that time, a formal authorization only); moreover, stakeholder role continues to be confined to the communication management processes, i.e., specifically, as it was in previous Guide, to *Communication Planning* process, which in any case was considered as a not-core facilitating planning process, and in *Information Distribution* process.

The third edition of *PMBOK Guide* (Project Management Institute, 2004) introduced a step forward in terms of processes, since, for the first time, there was a process that incorporated the word "stakeholder"; in fact, "*Manage Stakeholders*" process, which was supposed to "*manage communications to satisfy the needs of, and resolve issues with, project stakeholders*", was included in the communication processes, and it replaced previous "*Administrative Closure*", that was the process for "generating, gathering, and disseminating information to formalize phase or project completion", and which migrated, including some proper fortification, in its natural location of project closing processes. At long last stakeholders entered directly in project management processes, although focus continued to be limited to communications management processes, which, may be with a certain overvaluation, had also the not obvious (at all) task of satisfying the needs of, and resolving issue with, project stakeholders (what about managing deliverables? and changes?). In any case, stakeholder did not become a knowledge area yet, and the process of managing stakeholders was still considered as a process of communication management, as well as there were no news at all in terms of stakeholder identification, although, on the one hand, basic concept of *project charter* was added, and, on the other hand, interesting outputs of *Managing Stakeholders,* as *Resolved Issues, Approved Change Requests, Approved Corrective Actions, Organizational Process Assets Updates,* and *Project Management Plan Updates*, were included.

The following *PMBOK Guide*, the fourth edition (Project Management Institute, 2008), was of basic importance for stakeholder processes, since, although still staying in Project Communications Management, the very important processes "*Identify Stakeholders*" and "*Collect Requirements*" were added, and "*Manage Stakeholder Expectations*" replaced previous "*Manage Stakeholders*" process; but, in any case, in terms of stakeholder role, as we will see in the following ... all that glitters is not gold. *Identify Stakeholders* was defined as "the process of identifying all people or organizations impacted by the project, and documenting relevant information regarding their interests, involvement, and impact on project success", it was properly allocated in the Initiating process group, and it had as outputs both the *Stakeholder Register* and the *Stakeholder Management Strategy. Collect Requirements* was defined as "the process of defining and documenting stakeholders' needs to meet the project objectives", and the decision was to allocate this process in the Planning process group (of course in the scope, and not in the communications, knowledge area); although above definition of the process seems not to be completely clear, it seems to reflect, in this case too, the perspective of *requirement objectivity*, and a first key question may arise; *which came first, the stakeholders' needs or the project objectives? Manage Stakeholder Expectations*, which replaced previous *Manage Stakeholders*, was defined as "the process of communicating and working with stakeholders to meet their needs and addressing issues as they occur". In this definition, the word "working" seemed to be supportive in terms of developing effective relations with stakeholders, but, going in further details, it was specified that "*Manage Stakeholder Expectations involves communication activities directed toward project stakeholders to influence their expectations [...], such as actively managing the expectations of stakeholders to increase the likelihood of project acceptance by negotiating and influencing their desires to achieve and maintain the project goals*". Then, while above statement is perfectly clear if applied to generic "desires", another key question may arise: *are we sure that it is really possible and/or convenient to influence the stakeholder expectations, since, on one hand, the same expectations were foundational for the existence of the project itself, and, on the other hand, the satisfaction of the same expectations is the basis for achieving project goals?* In any case, *Managing Stakeholder Expectations* process had specific and accurate outputs, as *Organizational Process Assets Updates, Change Requests, Project Management Plan Updates,* and *Project Document Updates.*

Then, finally, in 2012, the turning point showed up: *the new International Standard "ISO 21500—*Guidance on project management/Lignes directrices sur le management de projet" (International Organization for Standardization, 2012)*—enshrined stakeholders' primary role.* ISO 21500, which was published after about a 5-year work of ISO Project Committee 36 (that included experts from almost 40 participating countries), had, and has (since it is presently still valid as international standard in project management), the purpose of providing high-level description (the document altogether consists of 36 pages only!) of concepts and processes that were considered to form good practices in project management. Indeed, *"Stakeholder" was defined as a "Subject Group"*, and other subject groups included *Integration, Scope, Resource, Time, Cost, Risk, Quality, Procurement, and Communication* ("*subject groups*" in ISO 21500 correspond, in through and through, to "*knowledge areas*" of PMBOK guides): the definition of the *Stakeholder Subject Group* is "*the stakeholder subject group includes the processes required to identify and manage the project sponsor, customers, and other stakeholders*".

Moreover, in ISO 21500, there are 2 of the 39 main processes (see Figure 2.1) that are directly related to stakeholder, and both of them are of course included in Stakeholder subject group; *Identify Stakeholders*, which is part of Initiating process group, and *Manage Stakeholders*, which is part of Execution process group.

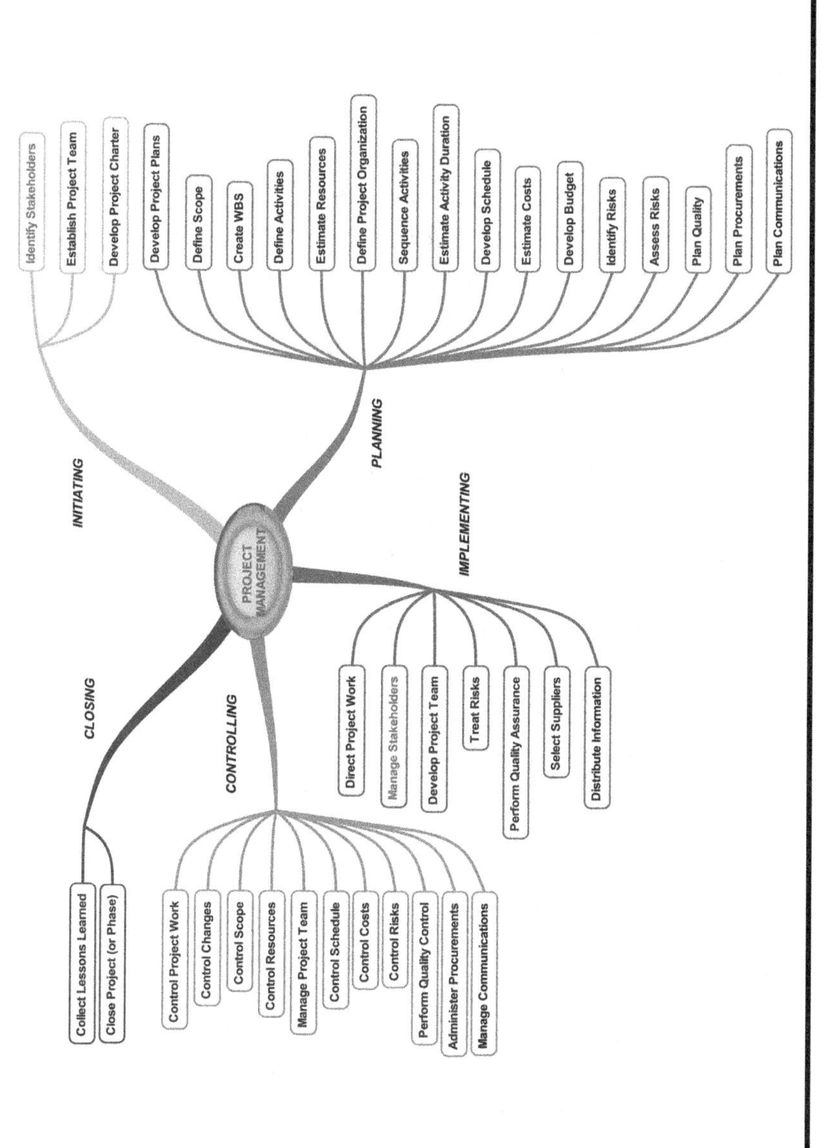

Figure 2.1 Mind map of ISO 21500 project management processes.

The purpose of *Identify Stakeholders* process is "to determine the individuals, groups or organizations affected by, or affecting, the project and to document relevant information regarding their interest and involvement", and its primary output is, then, the *Stakeholder Register.* As far as *Manage Stakeholders* process is concerned, its purpose is "*to give appropriate understanding and attention to stakeholders' needs and expectations*", while main output of this process is the *unsolved issues*, which originate *Change Requests*; then, *from now on, the stakeholder relations become central in project management.* Moreover, the utility of *soft skills*, which in the previous literature was considered limited to team development issues, spreads to all the domain of stakeholder relations; for instance, this is the first time that "new" concepts, as *diplomacy* and *tact*, are defined "*essential when negotiating with stakeholders*". Furthermore, we should notice the importance of the fact that, in *any perspective, the majority, or even all, of* Project Management *processes are, either directly or indirectly, related to stakeholder relations, too* (see Figure 2.1). In fact, besides the importance of the two main processes that are directly related to stakeholders, stakeholder relations influence deeply all initiating, implementing, and closing processes, but also a major part of planning and control processes; the remaining part of planning and control processes, which is mainly relevant to time and cost subject groups, could have, in theory, the project manager as its single actor, but, in the practice of today's world, it may be absolutely appropriate to engage the project team, too.

While *PMBOK Guide*, fourth edition (Project Management Institute, 2008), was basic for ISO 21500 realization, *PMBOK Guide*, fifth edition (Project Management Institute, 2013), which was intentionally released about one year later, *ensured alignment and harmonization with ISO 21500* itself. Moreover, the point that this last Guide "*aligns better with the focus on stakeholder management being put forward with the new ISO 21500 standard*" was considered as a strength;

after almost twenty years, stakeholder becomes a brand new knowledge area, and this confirms the acknowledgement of centrality, in Project Management discipline, of both stakeholders and stakeholder management. Project Stakeholder Management processes include both the "renewed" processes *Identify Stakeholders* and *Manage Stakeholder Engagement*, which were somehow present in the Project Communications Management Knowledge Area of previous fourth release, and the "brand-new" processes *Plan Stakeholder Management* and *Control Stakeholder Management*; from now on, the processes that are directly related to stakeholders are present in the Initiating, Planning (new), Executing, and Monitoring/ Controlling (new) process groups. *Identify Stakeholders* belongs to the Initiating process group, and it is defined as "*the process of identifying the people, groups, or organizations that could impact or be impacted by a decision, activity, or outcome of the project; and analyzing and documenting relevant information regarding their interests, involvement, interdependencies, influence, and potential impact on project success*". This definition enriches and strengthens previous one, and introduces a basic issue, i.e., the *stakeholder analysis*; the output of this process is still, of course, the *Stakeholder Register. Plan Stakeholder Management* is a new process, which is included in the Planning process group, and it is "*the process of developing appropriate management strategies to effectively engage stakeholders throughout the project life cycle, based on the analysis of their needs, interests, and potential impact on project success*". *This process is of basic importance*, not only *because it is the first time that a stakeholder-oriented process becomes part of planning*, but also *because*, together with the other two following processes, *stakeholder management comes out of communication management domain, and enters, by targeting the importance of effective engagement, in the domain of all-round relationship management.* Main output of this process is the *Stakeholder Management Plan*, which *now is in addition to the Communications Management*

Plan. The process *Manage Stakeholder Engagement*, which is included in the Executing process group, integrates previous Manage Stakeholders Expectation by *focusing on Stakeholders Engagement*, and is defined as "*the process of communicating and working with stakeholders to meet their needs/expectations, address issues as they occur, and foster appropriate stakeholder engagement in project activities throughout the project life cycle*". *From now on, in* PMBOK Guide, *Stakeholder Management integrates Communications Management,* and this is furthermore evident from the fact that *basic inputs to this process are both Stakeholder Management Plan and Communications Management Plan;* a major output continues to be *The Change Requests*. The *Control Stakeholder Engagement* process is the *first stakeholder-related process that has been included in the Monitoring and Controlling process group*, and it is defined as "*the process of monitoring overall project stakeholder relationships and adjusting strategies and plans for engaging stakeholders*"; as per other processes which are part of the Monitoring and Controlling process group, main outputs include Work Performance Information and, again, Change Requests.

After a while, also the other major International Association *IPMA, enshrined, coherently to ISO 21500, stakeholder role* (International Project Management Association, 2015). *Stakeholders, in fact, are a Practice Competence Element*, which "*includes identifying, analyzing, engaging and managing the attitudes and expectations of all relevant stakeholders*"; *the Competence Area Practice deals with core project competences, which include Project design, Requirements and objectives, Scope, Time, Organisation and information, Quality, Finance, Resources, Procurement, Plan and control, Risk and opportunities, and, precisely, Stakeholder*. Moreover, the purpose of the Competence Element Stakeholder "*is to enable the individual to manage stakeholder interests, influence and expectations, to engage stakeholders and effectively manage their expectations*"; then, *in ICB (Individual Competence*

Baseline) 4.0, there is a quite modern focus on stakeholder expectations, too. Furthermore, *stakeholders are obviously the core of the other Competence Area People, which deals with the personal and social competences of the individual,* including self-reflection and self-management, personal integrity and reliability, personal communication, relationships and engagement, leadership, teamwork, conflict and crisis, resourcefulness, negotiation, and results orientation, *and also of other Competence Area Perspective, which deals with the context of a project,* including strategy, governance, structure and processes, compliance, standards and regulations, power and interests, and culture and values.

Lastly, in PMBOK Guide, *sixth edition* (Project Management Institute, 2017), *the importance of stakeholders' role has been confirmed and enhanced*: "*Project Stakeholder Management includes the processes required to identify the people, groups, or organizations that could impact or be impacted by the project, to analyze stakeholder expectations and their impact on the project, and to develop appropriate management strategies for effectively engaging stakeholders in project decisions and execution*". The processes that relate to stakeholder management include

- *Identify Stakeholders,* almost identical to the correspondent one in previous guide, defined as "*the process of identifying project stakeholders regularly and analyzing and documenting relevant information regarding their interests, involvement, interdependencies, influence, and potential impact on project success*";
- the revised *Plan Stakeholder Engagement,* defined as "*the process of developing approaches to involve project stakeholders based on their needs, expectation, interests, and potential impact on the project*";
- the revised *Manage Stakeholder Engagement,* defined as "*the process of communicating and working with stakeholders to meet their needs and expectations, address*

issues, and foster appropriate stakeholder engagement involvement"; and

- *Monitor Stakeholder Engagement*, which replaced previous Control Stakeholder Engagement, defined as "*the process of monitoring project stakeholder relationships and tailoring strategies for engaging stakeholders through the modification of engagement strategies and plans*".

Definitively, today, standards recognize fully the centrality of stakeholders, but it seems that there is still a long way to pay the necessary attention to their needs and expectations, in order to satisfy them both. In fact, on the one hand, since maybe almost 90% of professionals have been certified in a context ante-*PMBOK Guide*, fifth edition, in which it was considered normal (and sufficient!) to give priority to hard skills to target project success, current phenomenon is still that the large majority of project managers, although it seems that they spend almost 90% of their time communicating with stakeholders, find it hard to recognize both stakeholder central role and the basic importance of soft skills. On the other hand, since the acknowledgment of stakeholder centrality is so recent, few stakeholder-dedicated literature and good practices exist, and this makes it difficult to implement adequately the stakeholder-relevant processes. *In any case, further steps in the direction of effectiveness with respect to both stakeholder identification and management seem necessary to increase the project success rate, especially in cases of large and/or complex projects*, and, therefore, some of these steps will be proposed in following chapters.

Chapter 3

Stakeholder Identification: Integrating Multiple Classification and Behavioral Models

The first process that is needed in order to develop an effective management of stakeholder relations is *Identify Stakeholders*, which is included in the Initiating Process Group, and, then, is preliminary to planning processes in all projects and in all of their eventual phases. The purpose of this process is to determine the individuals, groups, or organizations relevant to the project, i.e., affected by, or affecting, the project itself, to collect information regarding their interests, their needs and expectations, their influence, their involvement, their potential, their interdependencies, their impact on project success, and to document everything properly, e.g., in a stakeholder register. It should be noted that it is extremely appropriate to repeat this process of stakeholder identification *periodically*, since, in each project life cycle, stakeholders can quite often get in and out of the project domain, just as much as both their attributes and their relations can change.

Since stakeholders are numerous, are diverse, and have different needs and expectations, their identification has necessarily to be *analytical.* In fact, *on the one hand, their centrality requires a deep attention, and on the other hand, all stakeholders are important for targeting project success, while, in any case, forgetting a stakeholder who could be, or become, key would be an unacceptable risk.*

On the other side, project stakeholder domain is characterized by a *multilevel complexity*:

- *Stakeholders are persons*, or groups of persons, and we can assume that *persons are the most complex systems that exist* in the world.
- *Stakeholders are diverse*, and they are diverse *from different perspectives.* Project stakeholders, in fact, may have diverse interests and/or influence, may participate to and/or support diversely the project, they may add diverse value (either positive, or null, or negative), they may belong to diverse organizations and/or communities, and, since each organization is generally characterized also by a common business and/or social language, they may even speak or understand diverse organizational languages.
- *Stakeholders are numerous, and stakeholder relations are even more numerous.* If we consider stakeholders individually, in each project it could be very easy to distinguish hundreds of relations or thousands if we include the people from the web.
- *Stakeholder relations are context sensitive.* Indeed, both internal and external environments, just like business, social, or technological strategies, as well as economic, time, regulatory, legal, or social requirements and/or constraints, impact continuously on stakeholder relations.

- *Stakeholder relations may influence each other,* and this can happen continuously. Then, *taking care of both the relation with stakeholders and the relations among stakeholders becomes essential.*
- *All stakeholder relations,* due to their centrality, are important, and, at least, they *have to be monitored.*
- *Stakeholder relations may be evolutive in the life cycle of the project.* New stakeholders may come in, existing stakeholders may come out, each stakeholder may change level of importance and/or behavior, and this may happen several times.

Therefore, relations with, and among, stakeholders, introduce multilevel complexity in all projects: *classification models of stakeholders are mandatory to reduce that huge complexity, and, then, to make stakeholder relations issues addressable and manageable.* We can consider two main types of classification models for stakeholders, which are based on two diverse perspectives: *multiple classification models, which consider the belonging of stakeholders to different subjective categories,* and *the classification of stakeholders in communities, which reflects the stakeholder objective behavior.* Most commonly used multiple classification models are the grids, and especially the *power/influence grid,* the *stakeholder cube,* and the *salience model.*

The basic concept of the grids is categorizing stakeholders based on two of their main attributes, and then representing the results on a two-dimensional matrix. The most common grid is the *power/interest* grid (Mendelow, 1991), which categorizes stakeholders according to their level of authority in the project and their level of interest toward the project results, but also the *power/influence* grid and the *influence/impact* grid are present in the literature. Although power/influence grid has been developed almost 30 years ago, and it was created to be applied generically to organizations, rather than specifically to projects, it still has all its validity; in fact, with some

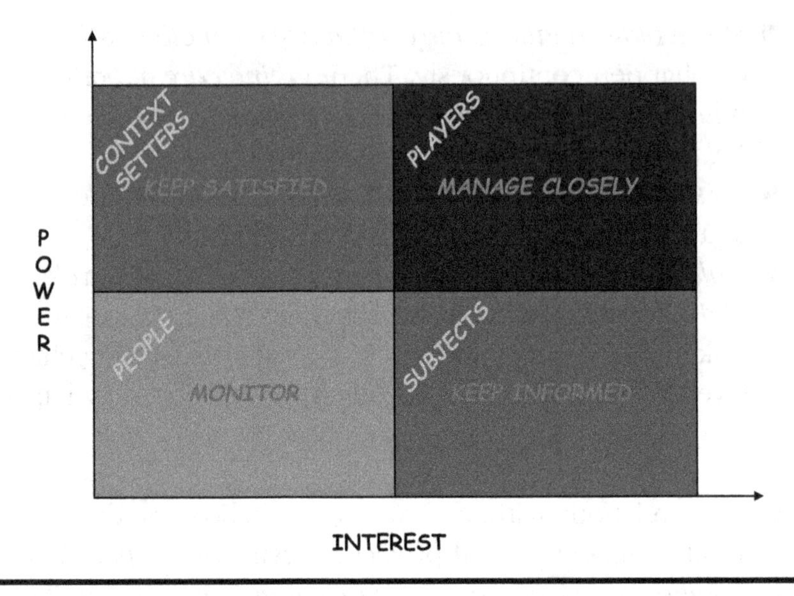

Figure 3.1 Power/interest grid.

customization, it is a tool quite simple to use, but immediately shareable, and enough powerful to support stakeholder identification, especially by categorizing them based on their importance for, and in, the project (see Figure 3.1, Power/interest Grid). In addition, complexity of actions to be taken may be drastically reduced by grouping the management of relations in four typologies: *monitor, keep informed, keep satisfied,* and *manage closely.*

Players include *key stakeholders*, as project team, project sponsor, top management, customers, users, and investors. Evidently, the relations with them *must be managed closely and generally, require direct, interpersonal communications.* Context setters generally include government, central and local public administrations, but may also include associations, trade unions, and, generally, all the organizations which establish constraints to the project in the form of laws, regulations, norms, standards, etc. *Relations with context setter stakeholders shall keep them satisfied*, and require compliance with the rules. *Subjects* may include interest groups, but also internal stakeholders, as functional managers and employees, who could *support* the project, so it can be also useful and convenient to *keep them informed*. Finally, *monitoring* the relations with/among other people and/or organizations involved, or that would like to be involved, generally requires the minimum level of effort, but it i*s even necessary*, not only because all the stakeholders are important, but also because, in general, stakeholders who are characterized by a certain level of power and/or influence could migrate, during project life cycle, to higher levels.

Stakeholder cube is a three-dimensional grid that maps stakeholders by introducing the third dimension of *Attitude*, in addition to the previous ones *Power* and *Interest* (Murray-Webster and Simon, 2006); while power is considered as the ability to influence the project, the level of interest states whether stakeholders will be active or passive, and attitude to the project indicates if stakeholders will either support the project or resist to it. Above three dimensions interact in eight different ways (see Figure 3.2):

- Saviours are powerful, active, have high interest, positive attitude, and managing closely the relations with them is necessary.
- Friends are low power, have high interest, and positive attitude, and they should be engaged as supporters.
- Saboteurs are powerful, have high interest, and negative attitude, and they need to be engaged in order to disengage.
- Irritants are low power, have high interest, and negative attitude, and they need to be engaged so that they stop "eating away".
- Sleeping giants are powerful, have low interest, and positive attitude, and they need to be engaged in order to awaken them.
- Acquaintances are low power, have low interest, but positive attitude, and they need to be kept informed.

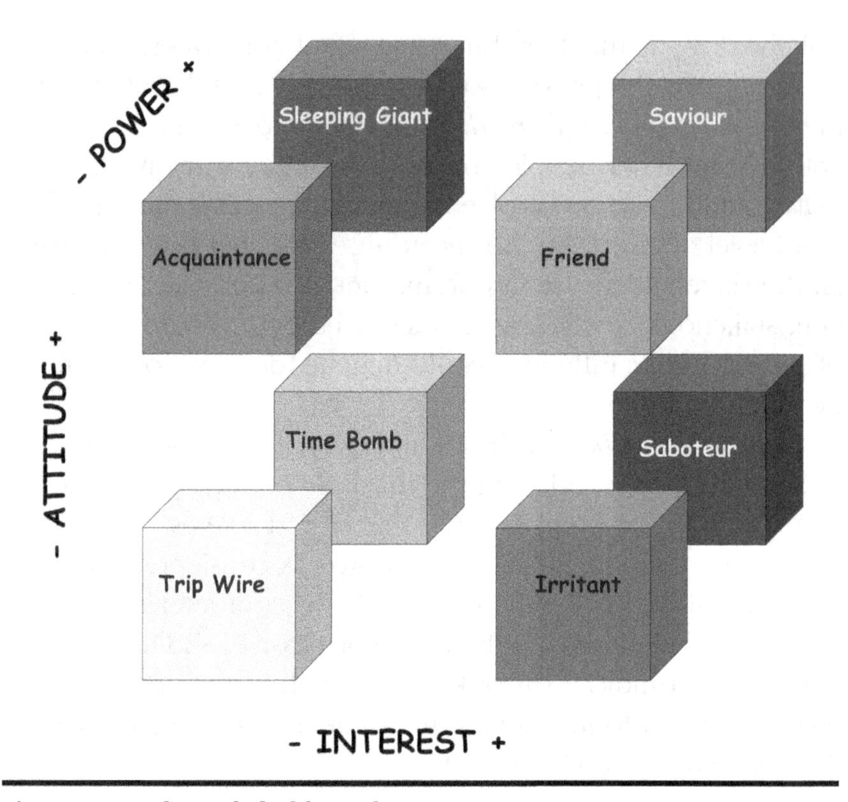

Figure 3.2 The stakeholder cube.

- Time bombs are powerful, have low interest, and negative attitude, and they have to be managed to avoid that "bomb detonates".
- Trip wires are low power, have low interest, and they need to be understood "watch the step".

Another multiple classification that is present in the literature is *Salience Model* (Mitchell, Agle and Wood, 1997), in which three main stakeholder attributes are defined, i.e., *Power, Legitimacy, and Urgency,* and then combined to identify each stakeholder group based on its "salience". In any case, since stakeholder attributes are variable (not steady state), socially constructed, and subjective (not objective), and, moreover, consciousness and willful exercise may or may not be present, in general, stakeholders may gain or

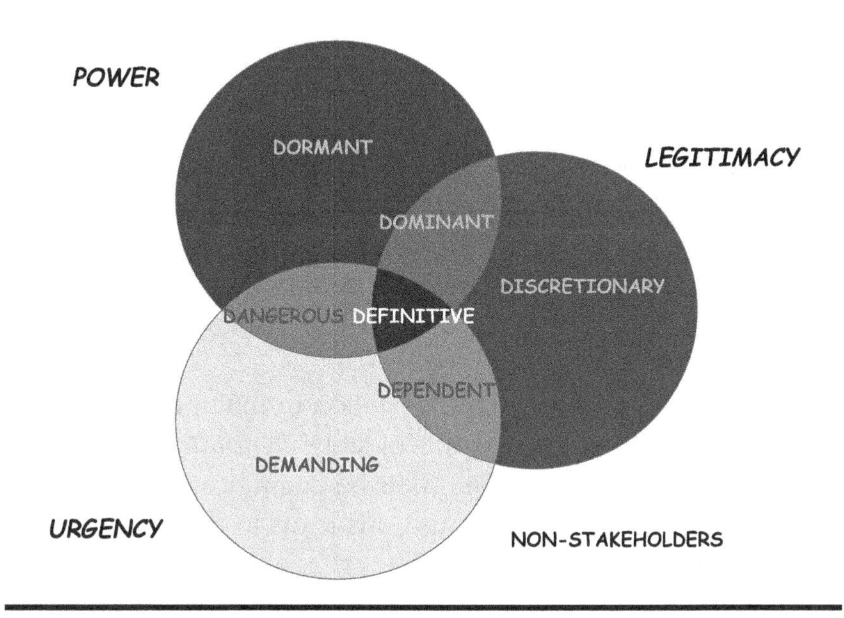

Figure 3.3 The salience model.

lose their salience during the whole project life cycle. In Salience Model, there are seven stakeholder typologies, either *Latent (Dormant, Discretionary,* and *Demanding)* or *Expectant (Dominant, Dependent, Dangerous,* and *Definitive)*, each one having either one, or two, or three attributes, plus the *Non-stakeholder typology*, which, of course, has no attributes at all (see Figure 3.3).

Therefore, in salience model, stakeholders are categorized based on their power, i.e., their ability to impose their will, their urgency, i.e., their need of immediate attention, and the legitimacy of their involvement.

Ultimately, multiple classification models are *quite effective* to prioritize stakeholders in accordance with their importance, and, specifically, to identify those *key stakeholders*, with whom, as we will see later on, *a direct relationship* is necessary to be developed; all of these models are based on the stakeholder *belonging* to certain categories. However, multiple classification models still leave some open issues that it is better to face, in order to increase both stakeholder identification efficacy and its effective usability by other stakeholder management processes, which are as follows:

- *While all the stakeholders are important, since they are central toward the project,* stakeholder characterization in multiple classification models is a *subjective process,* and the *importance of some stakeholder could be either over valuated, or, even worse, under evaluated, or ignored at all.*
- Project stakeholder characterization in multiple classification models is also, unavoidably, *dynamic,* because stakeholders may change their belonging category during all the project life cycle, and this leads to the need of a *continuous monitoring.*
- While in each project, *stakeholder behavior* can significantly either influence or being influenced by time, cost, and quality, *stakeholder belonging* to a certain category in multiple classification models per se does not.
- In multiple classification models, no correlations between categories and stakeholder expectations are evident, and/or specific, for each category.
- In multiple classification models, project stakeholders *maintain their individual behavior,* even if they belong to the same category, and/or they are at the same level of importance, and this *does not lead to a further reduction of the complexity* in following stakeholder-related processes, where *specific* actions have to be addressed.

Definitively, while stakeholder belonging to a certain category of importance is subjective, dynamic, not correlated with time, cost, and quality, and appears quite fragmentary in stakeholder domain, *categories that are based on stakeholder common behaviors and main interests are objective, durable, homogeneous, and directly related to project characteristics.*

The *behavioral* classification of stakeholders in *communities,* each one sharing a common prevalent interest and a common organizational language (Pirozzi, 2017), *is indeed a segmentation of the domain of stakeholders that helps effectively to reduce drastically the complexity of stakeholder management*, since it categorizes the whole domain of project stakeholders in four communities only. Moreover, since each community targets the prevailing variable/s quality/time/cost from the point of view of its specific interests, *there is the reliable advantage of a specific, accurate, and direct correlation among each stakeholder community and the three main dimensions that characterize each project.* Although this behavioral model is quite recent, there already have been major representatives of project management community who were so kind to support it (Archibald, 2017 and 2018; Stretton, October and December 2018) with their positive commentaries, then encouraging the author (myself) to proceed further.

In each project, there are, indeed, four main communities of stakeholders, which can be defined, respectively, as the *Providers*, the *Purchasers*, the *Investors*, and the *Influencers. Each one of these communities shares a prevailing interest in the project and a specific organizational language, and, then, stakeholders that are part of each of these categories have a common behavior.* Since each one of the four communities can be characterized by three main dimensions, and it is therefore representable with a cube, as well as the triad of variable quality/time/cost can be represented with a cube, too, the four communities of project stakeholders and the triad of variable quality/time/cost can be represented with a *hypercube* in 4 + 1 = 5 dimensions (see Figure 3.4).

The prevailing interest of the Providers is in the project as a whole. The project manager, the project team, the project management office, the suppliers, and/or the business partners share the common interest of realizing the project, in its optimal combination of the three main variables, i.e., time, cost, and quality. *Their specific organizational language is the*

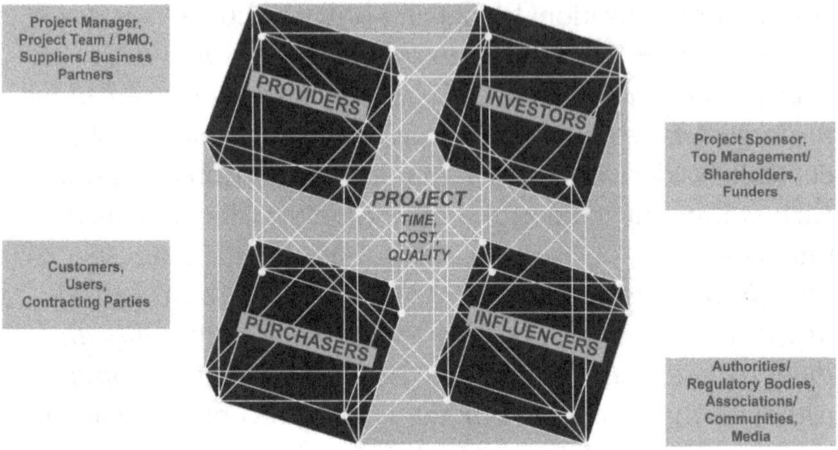

Figure 3.4 The stakeholder hypercube.

language of project management discipline, and their primary objective is the project completion within the triple constraints.

The prevailing interest of the Purchasers is the quality of the project. Both the customers, who contracted the project, and the end users, who will be the beneficiaries of the products/services that will be delivered by the project, focus on the common interest of obtaining from the project as much quality as possible, and this is also because they usually feel that both the costs and the time of the project are not further negotiable, while quality is. *Their specific organizational language is their business language, and the project for them is not a goal, but just a medium to achieve their own business goals.*

The prevailing interest of the Investors is the profitability. This result can be achieved by minimizing the project costs, by receiving the contracted income, and, possibly, by developing additional revenues over time; and it is only in this perspective of any business prospects that the delivered quality of the project could interest them. In the community of Investors, the sponsor and the project governance, the shareholders and the top management, as well as any external funder, are generally included. *Their specific language is the language of business economics, and the project is considered by them, in this case too, a medium to achieve their business goals.*

Finally, *the prevailing interest of the Influencers is to participate in the project*, even if they may not be a contracting party. In the community of the Influencers there are the authorities, such as the public administrations, the media, plus a large variety of other communities, e.g., the local communities, the lobbies, the trade unions, the associations, and so forth, as well as the negative/hostile stakeholders, such as the competitors, just like the personal stakeholders, and also that potentially very important group for project, and/or program, and/or portfolio, which is the domain of the potential customers and/or users. *Their specific language is the language of the media and/or the natural language, but sometimes and/or occasionally business language and the language of economics may be present too (Stretton, October 2018), while the project is for Influencers a medium that supports their goals and/or their own mission.*

Definitively, integrating a multiple classification model, as the power/interest grid, with the behavioral model of communities, can be very simple, since it is sufficient to associate to every stakeholder a letter that corresponds to each community (see Figure 3.5), and it is also effective at all, both to identify those key stakeholders to develop a direct communication with and to drastically reduce complexity of stakeholder management.

Finally, in project management, the major output of *Identify Stakeholder* process is the *Stakeholder Register*. This document contains information about identified stakeholders, which includes the following (Project Management Institute, 2017):

- *Identification information.* Name, organizational position, location and contact details, and role on the project;
- *Assessment information.* Major requirements, expectations, potential for influencing project outcomes, and the

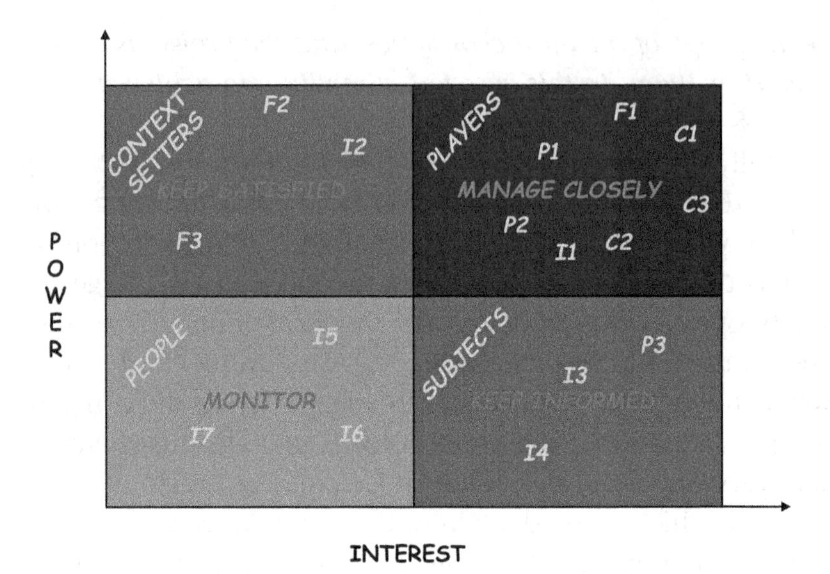

P = Providers, C = Purchasers, F = Investors, I = Influencers

Figure 3.5 The power/interest grid including categorization in communities.

phase of the project life cycle where the stakeholder has the most influence or impact; and

- *Stakeholder classification.* Internal/external, impact/influence/power/interest, upward/downward/outward/sideward, or any other classification model chosen by the project manager (e.g., the behavioral model in stakeholder communities).

Even if the first version of the document is an output of the Initiating Process Group, *Stakeholder Register is updated during whole project life cycle*, and, then, requires a specific, and delicate, management.

Moreover, since *Stakeholder Register is an input or an output of about the half of all project management processes*, it is a document that is vital for the project and that must be created carefully, and contents of which must be both accurate and characterized by a reasonable level of detail. While some information about stakeholders and their classification are a natural, and quite evident, result of stakeholder identification process, other *essential information*, including *diverse stakeholder expectations* and stakeholder *real role in the project* are normally *unknown and/or hidden*; this is why *stakeholder analysis is fundamental* for achieving project success, too.

Chapter 4

Effective Stakeholder Analysis: A Systemic Approach

An effective stakeholder analysis is essential to target project success, because:

- *stakeholder expectations are different, and they have to be harmonized and/or prioritized via a decision process;*
- *the adherence of project requirements and constraints to stakeholder expectations must be verified and validated;* and
- *some important stakeholder expectations, and even some stakeholder roles in the project, may be unclear and/or hidden, and have to emerge properly.*

Stakeholder analysis is presently described, in project management literature, either as *a part of Identify Stakeholders* process "a detailed analysis should be made of stakeholders and of the impacts they might have on the project, so that the project manager can take maximum advantage of their contribution to the project" (International Organization for Standardization, 2012) or as *one of the "Tools and Techniques"*, which is present in each of the *Plan Risk Management, Identify Stakeholders*, and *Monitor Stakeholder Engagement* processes (Project Management Institute, 2017). In this latter case, stakeholder analysis is supposed to result in a list of stakeholders which include basic information such as their expectations, their positions in the organization, their roles on the project, their levels of support of the project, and their "stakes", i.e., their interests, rights, knowledge, ownership, contribution. Whatever the approach is, *an effective stakeholder analysis is foundational to define properly project scope*, which is one of the basis of possible project success, since the first question we have always to answer to in all projects is "what do we *really* have to do?".

Which is the real situation today in terms of understanding stakeholder expectations and properly assessing project scopes? Although it seems that our project management community does not like to talk very much about this topic, it is evident from valuable PMI's surveys (Project Management Institute, 2018) that projects do not seem to be so successful investments, *since, today, more than 30% of projects do not meet their original goals and business intent, i.e., they do not satisfy stakeholder expectations, and almost 50% of the projects experience scope creeps* (clearly, a little lower percentages are valid also for projects that are not completed within their initial budget, and/or within their initially scheduled times), *i.e., their results do not match anymore with initial project requirements.* This situation is so common that one of the greatest authors in project management, Harold Kerzner, states "there are three things that most project managers know will happen with almost certainty: death, taxes, and scope creep …. Scope creep is a natural occurrence for project managers. We must accept the fact that this will happen" (Kerzner, 2017). Main causes of above effects include, besides of course *project complexity*, the following:

- *misunderstanding, and/or lack of knowledge, of stakeholder expectations*;

- *existence of unsolved conflicts among diverse stakeholder expectations*;
- *unawareness of constraints of all types* (e.g., legal, situational, environmental, organizational, business, social, etc.);
- *poor understanding and/or definition of project/stakeholder requirements*; and
- *presence of discrepancies and/or inconsistencies between stakeholder expectations and project/stakeholder requirements.*

It is evident that, in all the above cases, an effective stakeholder analysis could significantly improve the situation in terms of accuracy of project scope, care of stakeholder expectations, and, ultimately, definition of a viable path toward project success.

Since stakeholder expectations are present in each of the investment, project, and product life cycles, while stakeholder requirements are present in project life cycle only, the *first step of a systemic approach to effective stakeholder analysis is to determine properly the different cause/effect relations in the whole investment life cycle*, and, moreover, the different actors and relevant perspectives that influence the project scenario (see Figure 4.1).

In the first investment phase, there are three different perspectives, which are the Investors, the Purchasers,

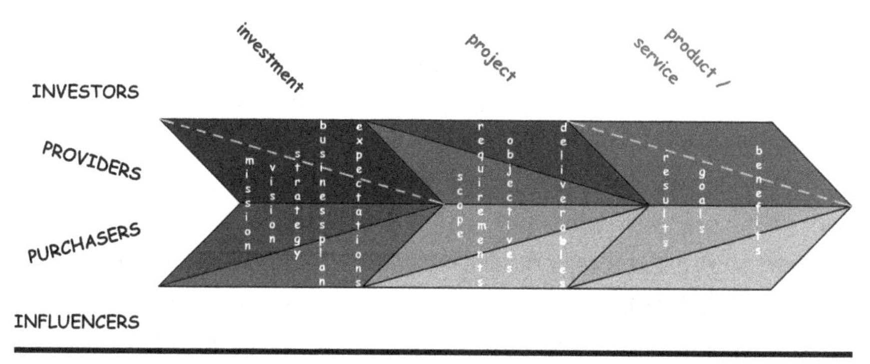

Figure 4.1 Different perspectives in project investment value chain.

and the Influencers: of course, since the Providers will just be appointed in the following project phase, in this phase they can be considered only as an undistinguished part of Investors' domain. *All three above-mentioned stakeholder communities define their own strategies in accordance with its mission and vision, develop their business/social/other plans, to realize their own strategies, and, then, define their own expectations which come out from their plans: at this stage, if supply and demand match, Purchasers and Investors communities bring together their diverse, but considered as compatible, expectations, formalize their agreement, in a contract or similar, so that project, with its requirements and its constraints, comes alive.*

This passage is crucial for future project success: *diverse "subjective" stakeholder expectations are somehow harmonized in "objective" project requirements and constraints. While the mediation between Purchasers' and Investors' expectations generate both project requirements and the triple constraint, i.e., states time-cost-quality in project scope, disparate Influencers' expectations should either generate other constraints (legal, normative, business, etc.), or, if insoluble conflicts are present, lead to prioritizing decisions that have to be mutually agreed between Investors and Purchasers. If, here in this passage, there are either mismatches, or misunderstandings, or incompleteness, or inaccuracies, or mistakes of any type, which are not properly solved via an efficient stakeholder analysis, and, then, managed via an effective stakeholder management, unavoidably, when project will be completed, there will be a certain degree of stakeholder dissatisfaction, and, therefore, of project failure.*

In project phase, indeed, as we saw previously, there are four different perspectives, which are those of Providers, Investors, Purchasers, and Influencers. *These perspectives are different, but from now on, they are supposed to "share" a common view in terms of scope, requirements, constraints, objectives, and deliverables.* In fact, *the project will be successful only if its results will satisfy stakeholder expectations, which are, on turn, inextricably linked to the stakeholder perception that, in following product/service phase, when perspectives will become three again, project goals will be achieved, and expected benefits will be obtained. On the other hand, during project phase, mismatches and/or misunderstandings about stakeholder expectations will lead in discrepancies between agreed project objectives and perceived project goals, so generating a high probability of project failure.*

An effective stakeholder analysis has the purpose of properly assessing stakeholder expectations, in accordance with their different perspectives: behavioral classification in communities, as per previous chapter, is very helpful, in this case too, to reduce complexity, and this usefulness is strengthened by the fact that the expectations of the different stakeholder communities target the product/service life cycle, and, then, tend to remain constant in project life cycle, exactly as it happens for their behaviors. For each stakeholder community, *the effective analysis can therefore be based on a systemic approach, which focuses on cause-effect relationships: in all cases, relationships between strategies, which are the causes, and the expectations, which are the effects, are defined in the business and/or other plans that stakeholders set up, and they are affected by the internal and external environment, too* (see Figure 4.2).

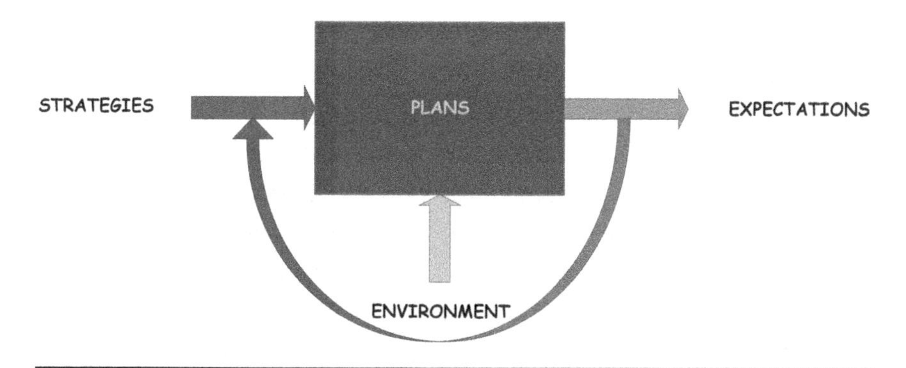

Figure 4.2 Cause/effect relation between strategies and stakeholder expectations.

In Investors' perspective, plans are generally business plans, and, generally, they are available and/or accessible through the project sponsor: Investors' economic and financial expectations rely on project, and on project follow up, and, then, influence directly project objectives and triple constraint.

In contrast to Investors' plans, those of Purchasers' generally are not available and accessible to project manager, and, then, they look unknown and/or hidden. An in-depth study of both customers and users business and/or social context, integrated by discussions and interactive clarifications with Purchasers, as well as, preferably, the development of some close personal and professional relationships with key stakeholders, become foundational to understand correctly their expectations, and, afterwards, to monitor them properly. Purchasers' expectations concern mainly the product/service life cycle and, then, impact directly on project goals, so their influence on project objectives is indirect, but essential; definitively, analyzing Purchasers' expectations is the hardest part of the work, but it is basic to target project success.

Ultimately, Influencers' plans are generally not evident too, as per those of Purchasers', but, since they are not directly involved in the contract, their influence in the project does not concern directly scope and objectives, although it proves to be basic in establishing constraints of different types (legal, regulatory, environmental, etc.).

Potential conflicts and/or misunderstandings between different stakeholder expectations must be immediately solved, and/or an agreed prioritization has to be made, just like initial scope and requirements have to be reviewed accordingly: indeed, only the alignment of the diverse stakeholder expectations can guarantee a proper project development, and, ultimately, an increase of the project success rate.

Definitively, in our today's world, it becomes evident that *project manager competencies must include the management of not only operational issues, but also of strategic issues, which, furthermore, have to concern not just the domain of his own organization, but even customers, users, and other Influencers' domains. A good understanding of these domains may be facilitated, and accelerated, by some effective relations with those key stakeholders, who, if properly engaged and self-engaged, could help a lot: a successful key stakeholder management, who can rely on an effective communication, becomes, then, a powerful and foundational mean to target properly project success.*

Chapter 5

Key Stakeholders Management: Principles of Effective Direct Communication

In all projects, relationships with key stakeholders, due to their importance and/or to their potential impacts on the project itself, have to be managed "closely" (see Chapter 3); these relationships must be then necessarily supported by "close" forms of *communications, effectiveness of which evidently requires an approach that is mostly not only interpersonal, but also direct. Above directness involves mainly two aspects: first one is the "individuality" of communication, meaning that communication has to be in the form of one-to-one, or one-to-few, while second one is the communication need of being not mediated.*

Meanwhile, more generally, it is essential to clarify the cause-effect link between stakeholder relations and communications. Indeed, *there is an unbreakable bond between* Project Stakeholder Management *and* Project Communications Management *processes, which are linked in a mutual support relation; effective stakeholder management requires effective*

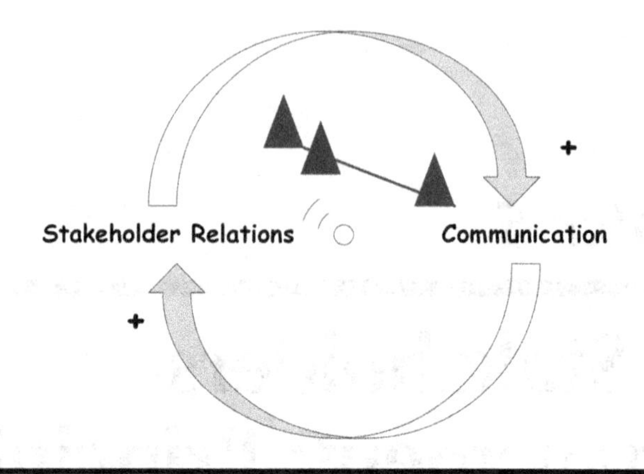

Figure 5.1 Reinforcing loop with "snowball effect".

communication, and effective communication is possible only if relations with stakeholder are very good. Furthermore, the importance of this relation is strengthened by the fact that, during the project development, there can be several moments in which *communications may constitute the only deliverable that can be actually delivered to key stakeholders.*

Indeed, *there is a virtuous circle that includes both stakeholder relations and communication,* which forms, in a *systems thinking* approach (Senge, 2006), a *reinforcing loop: a growth in stakeholder relations generates an increase of effective communication, and so on* (see Figure 5.1), although, vice versa, poor stakeholder relations generate inadequate communication, and so on. In fact, effective communications in stakeholder relations are basic for stakeholder satisfaction, and may be this is one of the principal reasons about the project managers and executives' convincement that *almost one-half of unsuccessful projects have ineffective communication as the main contributing factor* (Project Management Institute, 2013).

But, still in a systems thinking approach, what are the *"Limits to Growth"* (Senge, 2006) that have to be mitigated and/or removed in order to allow effective project stakeholder, and communications, management? *The main limit to growth that hampers the effective development of both stakeholder and communication management is the potential inadequacy of project manager competencies*. In fact, *project managers need not only specific brilliant soft skills in effective communication, but also*, since unavoidably the time spent in communication is 80 percent or more of their total working time, they need *those excellent hard skills that will enable above soft skills to be put into practice adequately, and for the proper time*.

Indeed, it seems that, presently, *there is a good awareness about the essential role of communications*, since, at first, four-fifths of project managers and executives believe that communication is today more important for project success than it was 5 years before (Project Management Institute, 2017), and, moreover, there are already some years that the lack or the inadequateness of project effective communication is considered one of the main causes of project deemed failures (Project Management Institute, 2013). Nevertheless, unfortunately, *for many years, and, specifically, up to when stakeholder became a subject group* (International Organization for Standardization, 2012), and *project stakeholder management became a knowledge area* (Project Management Institute, 2013), *project communication management was de facto considered, by the project management community, as "independent" from project stakeholder management*.

In fact, first definition of project communications management was that it "includes the processes required to ensure timely and appropriate generation, collection, dissemination, storage, and ultimate disposition of project information" (Project Management Institute, 1996), while current edition is a bit more complete, but, in any case, lightly different, since it states that "project communications management includes the processes required to ensure timely and appropriate planning, collection, creation, distribution, storage, retrieval, management, control, monitoring, and ultimate disposition of project information" (Project Management Institute, 2017). Both definitions, indeed, are evidently focused on "information management", rather than on "communication management". The main difference between communication and information is that, while *communication is a two or more way transfer that involve human relations aspects, information can be limited to a one-way rational transfer, which does not necessarily involve any feedback or existence of personal relation.* Maybe that this sort of "misunderstanding" was, and still is, a contributing factor to project inefficiencies and/or failures, since relationships were not considered always essential, as definitively it is proven they are, especially because they determine stakeholder satisfaction. Indeed, the etymology of word "communication" is from Latin "communicatio", roots of which indicate "a common participation", or "the action of letting someone in on a subject"; therefore, from ancient times, *the concept of communication is bound to the sharing of information*, rather than their unidirectional broadcasting, and, then, *the model of communication is not purely linear, but absolutely interactive.*

So, what is stakeholder communication? Basically, it is a crucial, and constantly present, aspect of *interaction among stakeholders*, and, then, it can be considered as *a synonym of behavior* "In the perspective of pragmatics, all behavior, not only speech, is communication, and all communication (even the communicational clues in an impersonal context) affects behavior" (Watzlawick, Beavin, and Jackson, 1967). Five axioms can help to describe properly the communication interactive process (Watzlawick, Beavin, and Jackson, 1967):

1. *One cannot not communicate.*
2. *Every communication has a content and a relationship aspect*, such that the latter classifies the former, and is therefore a metacommunication.
3. *The nature of a relationship is contingent upon the punctuation of the communicational sequences between the communicants.*
4. *Human beings communicate both digitally and analogically.*
5. *All communicational interchanges are either symmetrical or complementary*, depending on whether they are based on equality or difference.

The first axiom states that, since communication is a behavior, and it is impossible not to behave, even not communicating is, and is interpreted as, a specific communication strategy; moreover, in general, everything a person does is a communication "activity or inactivity, words or silence all have message value: they influence others and these others, in turn, cannot not respond to these communications and are thus themselves communicating" (Watzlawick, Beavin, and Jackson, 1967). In other words, and besides what might be thought, *communication exists in all cases, either it is intentional or not, conscious or not, purposeful or not, successful or not; the actual issue to deal with is, then, "not communicating" or not, since second alternative does not exist, but "communicating effectively".*

The second essential axiom states that *each communication is characterized both by a level of content and by a level of relationship*, and in fact it is precisely the relationship level that interprets the content level, so, ultimately, assessing which content is really perceived and "how this communication is to be taken", since "all such relationship statements are about one or several of the following assertions: this is how I see myself...this is how I see you...this is how I see you seeing me..." (Watzlawick, Beavin, and Jackson, 1967). Definitively, *the content aspect matches with the information, while the relationship aspect, which is a metacommunication since it classifies the content level, is "an information about information"; it is then evident that the relationship becomes essential to develop effective communication.*

The third axiom is concerned with the punctuations that participants include in their communication sequences, which correspond, and lead, to *interpretations that can be very different*; in fact, *punctuation organizes behavioral events, and, then, it can be considered essential to ongoing interactions.* Since, in any communication act, "every item in the sequence is simultaneously stimulus, response, and reinforcement" (Bateson and Jackson, 1964), each part of the communication sequence can be interpreted, by different participants, as either a stimulus, or a response, or a reinforcement, and, then, there is the risk that disagreement about punctuations becomes the cause of countless and/or endless relationship struggles.

The fourth axiom states *the importance of both verbal and nonverbal communications*, which are generally combined together, to target the development of relationships; communication is digital when we use words to represent things, while communication is analogic when we represent things via images and/or via some forms that are related to the other senses. To summarize, "digital language has a highly complex and powerful logical syntax but lacks adequate semantics in the field of relationship, while analogic language possesses the semantics but has no adequate syntax for the unambiguous definition of the nature of relationships" (Watzlawick, Beavin, and Jackson, 1967).

In the fifth axiom, communicative interactions are mapped in either symmetrical or complementary, if correspondent relationships are based, respectively, either on equality or on difference. In symmetric communication, each participant aims to mirror the behavior of the other, and both feel to be at the same "level", while, in complementary communication, each participant complements the behavior of the other, and four different reciprocal positions are possible, i.e., one-up, one-down, "metacomplementary" (one lets or forces the other to be in charge of him), and "pseudosymmetry", (one lets or forces the other to be symmetrical). Evidently, *there could be conflicts if either a participant does not accept and/or recognize his role, or a participant wishes to change the current situation.*

Therefore, it is evident that, fortunately, since several decades communication theory concentrate on the unique role of relationship, adaptations in a project context may be useful and appropriate; in fact, communications between key stakeholders can be represented with a *communication interactive model* (which is an adaptation of Schramm's model, 1955), where it is possible to highlight the main factors that have to be addressed in order to manage relations effectively (see Figure 5.2).

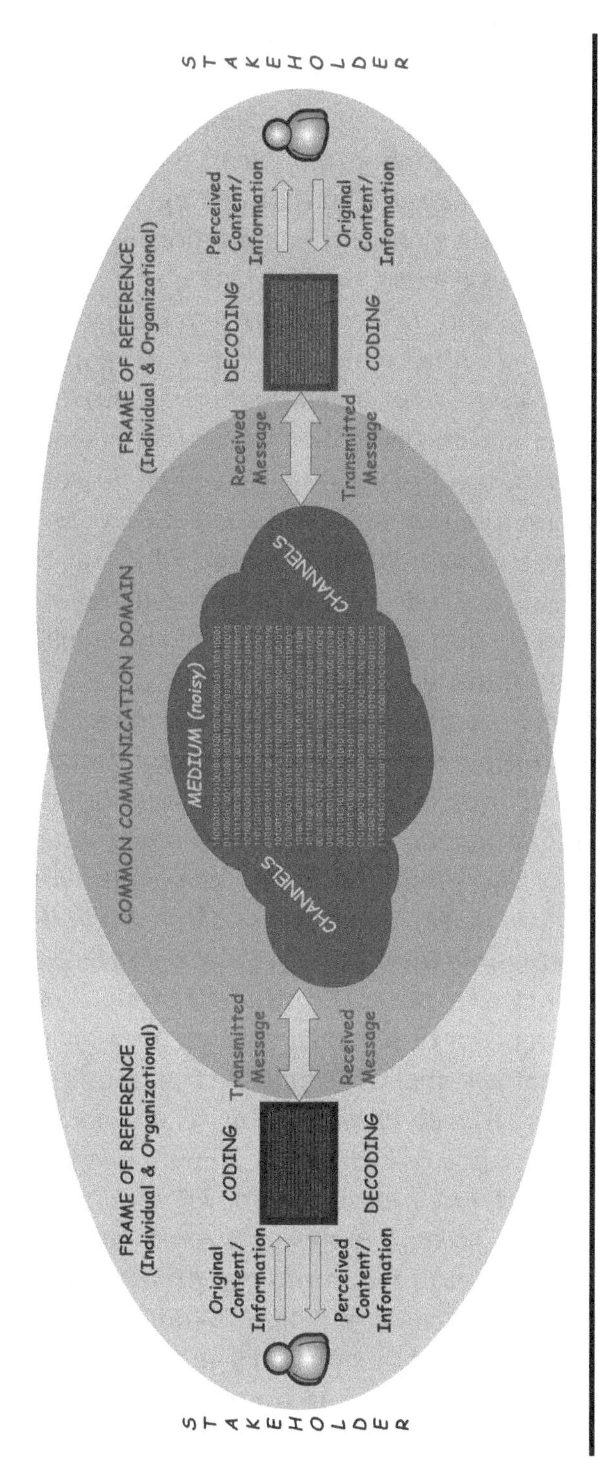

Figure 5.2 Interactive stakeholder communication model.

Communication flow is a continuum of communication sessions, each one influencing the following ones and including both contents, i.e., objective, and relationship, i.e., subjective, aspects, and in which each stakeholder has alternatively the role of sender and receiver. The transmitting stakeholder, who has his/her own individual and organizational *frame of reference, codes* original content or information in a specific language and format, and *delivers the correspondent message in a noisy medium/channel,* which is, on its side, part of a specific *communication environment that is necessarily common* between sender and receiver; then, receiving stakeholder, who, from his/her part, has his/her own individual and organizational frame of reference, gets the message, decodes it, and receives a *perceived content/information,* which, at this point, will influence his/her following behavior resulting in a new communication session in which he/she changes his/her role from receiver to transmitter, and so forth.

Stakeholders introduce human factors in the relationship process, and *transform objective information in subjective, either confirmed or distorted, communication, through their personal and organizational frame of reference,* which applies a *multiplicity of filters that may, at the same time, enhance, minimize, or even delete, diverse parts of the original content. In general, communication will be effective if stakeholder subjectivity will succeed to positively value the original informative contents.* The concept of personal and organizational frame of reference (that is somehow similar to Schramm's "field of experience") is that *each human being,* who is continuously bombed by thousands of elementary stimuli, *selects and categorizes, in order to manage and/or interpret the situation carefully, only those stimuli that he personally believes, and/or perceives, are important according to his own judgement.* Personal filters may include, but are not limited to, spoken language, emotional and/or situational state, cultural and/or historical and/or geographical and/or religious and/or personal beliefs and/or socio-economic and/or familiar and/or health context, concept of the self, cognitive skills, disturbs and distractions, fatigue, age, previous experiences, personal style, self-protection mode, etc., while organizational filters may include, but are not limited to, business and career languages, professional expectations and/or roles and/or levels of authority, leadership, hard and soft skills, etc.

The coding process transforms contents, which have been somehow filtered by personal and organizational frame of reference, *in messages*, which must have those characteristics that are suitable to be properly delivered to the communication channels and media, to make the stakeholder target reachable, and, at least theoretically, to give to him the capability, after a proper decoding process, of correctly perceiving, and, then, understanding, the message itself. *Coding and decoding processes are then essential to address a commonly reachable, and understandable, communication environment, and they can achieve this goal only if they use shareable languages and formats. A language, in order to be actually shareable, at first, of course, must be derived by a spoken language that can be understandable by both stakeholders, but, moreover, in order to avoid misunderstandings and/or to reduce eventual semantic noise, must contain also those elements that reflect the specific languages of the relevant stakeholder communities.* In fact, as we saw in Chapter 3, *each stakeholder community has its own language, in which the same community recognizes itself*; for each stakeholder community, the choices of common languages and formats are bold decisions, which correspond to deep bonds in terms of identification, affiliation, and efficient practice. *Ultimately, the communication language to be used, in order to be effective, should contain that combination of natural, project management, economic, business, social, bureaucratic, political, media, social media languages that correspond to the two organizations of both the transmitting and the receiving stakeholders.*

Messages are those formatted and coded communication items, which *are exchanged between stakeholders via the commonly available channels and media*. There are *informative messages*, which of course constitute the core of stakeholder communications, but there can be also other types of messages, which support the effectiveness of stakeholder communications: the *feedback messages*, which are very special answer messages that return information about the delivery and the receipt of sent messages, and management of which is of extraordinary importance for communication effectiveness; the *feedforward messages*, which have the purpose of anticipating at best messages to be sent; and the *metamessages*, which are "messages about messages" that have the purpose of a better tuning between end-to-end communications.

Furthermore, *each communication between stakeholders can exist if and only if there is a common communication domain*, which is the result of an intersection between the two diverse stakeholders' personal and organizational frames of reference, and *which contains both the channels and the media that are mutually available to both stakeholders who wish to communicate. Therefore, no communication is possible if a common domain either has not been established, or become inactive, or it has been closed by at least one of the communicating stakeholders. In fact, common communication domains between stakeholders do not necessarily exist at project start-up, and in this case, they have to be set up, as well as these domains have an absolute dynamical nature, so that they have to be maintained during whole project life cycle.*

Setting up a common communication domain, namely *opening a communication*, and maintaining live the same common domain, namely *keeping communication channels open*, are, in today's world, two issues that are characterized not only by a foundational importance, but also by an extraordinary sensitivity. In fact, *globalization enhances mass and impersonal communications much more than interpersonal communications,* and people, in both their personal and organizational domains, are generally reluctant and/or suspicious toward establishing new relationships, and far too careful in developing existing ones. In order to solve this essential, but complex, problem, it is better to be very pragmatic, and, then, to trace back that one of the core characteristics of communications between key stakeholders is, as we will see in more detail in the following, *purposefulness*; i.e., each of these communications has some purpose and, then, *corresponds to some expectations*. Therefore, an *effective stakeholder analysis*, which has as its results also the evidence of different expectations and of their harmonization, *is a powerful tool to set up common communication domains*. Furthermore, as far as maintaining common communication domains is concerned, we can trace back that, as previously seen in this chapter, *communication is behavior*, and, then, if we *focus on behavioral stakeholder classification in communities*, we have just four principal behaviors, i.e., the behaviors of Providers, Purchasers, Investors, and Influencers, respectively, which are almost constant in project life cycle. Indeed, considering the six intersections of these four frames of reference *can simplify a lot the maintaining of correspondent communication domains*, and, then, may let to concentrate properly on the *most effective dynamics of communication channels* to be used.

Selecting and using properly communication channels is foundational for each communication effectiveness; indeed, *above key tasks are the results of specific decisions that have to be made based on both stakeholder and environment analysis*, and are not either routine or common sense activities. Moreover, since relations with and among stakeholders are evolutionary too, reaching and maintaining effective communication requires also a *monitoring approach, which is based on active listening and observation, in order to address a continuous learning from the feedbacks*, and, more generally, *a dynamic Plan–Do–Check–Act approach is absolutely preferable.*

Each stakeholder manages and perceives any communication, both in terms of content and in terms of relationship, through his five senses, and, then, communication channels have to be designed, planned, implemented, controlled, and improved, accordingly. Naturally, the communication channels that are most suitable for developing an effective communication address are either *hearing*, or *sight*, or *a combination of both*, and may be pragmatically categorized in *oral, paraverbal, non-verbal, written, and visual communication*, while communication channels based on touch, taste, and smell are generally used not to develop, but mainly to support effective professional communications. Both *oral and written communications constitute verbal communication, since they are both based on the use of words ("verba" in Latin) to represent material and immaterial concepts.* As per fourth communication axiom (Watzlawick, Beavin, and Jackson, 1967), verbal communication represents the digital part of the communication, while nonverbal and paraverbal communication (which may integrate oral communication) and visual communication (which may integrate both oral and written communication) represent the analogical part of communication.

Oral communication is based on voice, and is made of conversations (from Latin verb "conversari", which means "stay together") between two or more stakeholders, who can *interact either in a face-to-face modality, generally integrated with nonverbal, paraverbal, and visual communications, or in a remote modality,* e.g., via call conferences, webinars, etc., *or via phone calls,* which are purely vocal, but which are still integrated with paraverbal communications. *Due to its nature,* which is basically that of being a channel very interactive, rich of information, and flexible, but volatile (Latin proverb states "verba volant, scripta manent", which means, "words come and go; only the written word remains"), *oral communication is mainly indicated to develop the relationship part of communication, and less indicated to agree contents.* In general, *main advantages of oral communication may be interactivity, flexibility, time and cost saving, possibility of getting feedbacks quickly, better capabilities of persuasion and control, while main disadvantages may include excess of emotions, lack of consideration and/or of legal validity, difficulty in assessing responsibilities, presence of confusing items and/or of less clarity, and unavailability of reliable records.* The large variety of oral communication channels includes speeches, meetings, presentations, seminars, webinars, talks, interviews, conferences, formal and/or informal lunches and/or dinners, visits, conventions, events, telephone calls, and conference calls; a further, peculiar, but very effective, communication channel is the *provision of training* (Gabassi, 2006).

A number of years have already passed since research showed (Mehrabian, 1971) that perception of face-to-face messages is driven just 7% by spoken words, while 38% is driven by paraverbal communication, and as many as 55% is driven by nonverbal communication (i.e., body language), as well as, in phone calls, where nonverbal communication goes obviously to zero, the importance of spoken words arises to 13%, but that of paraverbal communication increases up to 87%; *paraverbal and nonverbal communication are, therefore, someway dominant in interpersonal oral communications*, and this happens both in face-to-face and in purely vocal cases.

Paraverbal communication includes all those paralinguistic aspects, both conscious and unconscious, which can modify, enhance, and attenuate the meaning of an oral communication by influencing considerably its perception in the listeners. There are several paralinguistic aspects that accompany each vocal message, such as, for instance, the *intonation*, which qualifies each statement in terms of communication intentions, the *use of different speeds and pauses*, which are helpful to emphasize the diverse parts of oral communication, the *loudness*, which it is used not only to overcome distance, but also to characterize the communication, the *pitch*, the *timbre*, and also some specific forms of *paralinguistic respiration*, e.g., gasps, sighs, throat-clears, and "mhms".

While paraverbal communication focuses more on giving meanings to the messages, *nonverbal communication is mainly used to validate speeches*; in fact, although human beings, if compared with other species, developed an incredibly sophisticated communication system based on voice, the sense of sight is still preferred because of its better objective reliability. Paraverbal communication is often synonym of *body language* (Pease and Pease, 2004); *posture* (while sitting, standing, talking, listening, etc.), *facial expressions* (especially *eyes and mouth* movements, and their coherency), *gesture* (symbolic, to replace words, but also to show emphasis, satisfaction, stress, comfort, adaptation, anger, etc.—and their contraries), *proxemics* (effects due to distance, position, and man's use of the space—Hall, 1982), *dress code*, etc. *are extraordinary supports, both conscious and unconscious, of a vocal message, and contribute significantly to its effectiveness—or to its failure. Interpretation of body language can be basic also to understand if a speaker is lying.* In fact, since most of the people have a certain ancestral capability to perceive some body language signals, *it is highly risky to expose things that are far from reality, or to present "images" rather than the "selves", since listeners could catch some deviations and, then, refuse the whole communication.* Ultimately, in any case, it would be very important, for project managers, in order to improve the effectiveness of their communication, to have also some skills in body language, and, if possible, in NLP—Neuro Linear Programming (Bandler and Grindler, 1982) too.

Written communication is another verbal form of communication; due to its intrinsic solidity and stability, *written communication is particularly indicated to agree contents*, and it is less indicated to develop the relationship part of communication. In written communication, stakeholders interact, in *deferred time*, through the exchange of written messages, including emails, letters, all types of documents (reports, proposals, minutes, memos, etc.), and, although, in professional environment, with a quite limited use that is subordinated to other forms of communication, also by exchanging social media and mobile technology messages. *Written communication can be effectively integrated with visual communication*, since the association of words to images may have both a greater informative impact ("a picture is worth a thousand words"), and a better liking. In interpersonal communication, above-mentioned integration is common in all type of documents, in dashboards and scorecards, in slides for presentations, and, although this applies mainly in impersonal communications, in brochures, flyers, posters, and other type of printings; moreover, *an appropriate choice of the communication environment (location, meeting room, etc.) can be extraordinary helpful. Main advantages of written communication are its validity, its unobtrusiveness, its stability, the easiness of permanent recording, the possibility of producing it anytime and anywhere, while main disadvantages may be both time and cost that are required, the delay of feedbacks, the absence of flexibility, the difficulty of making changes, the substantial neutrality and "coldness" with respect to the relationship, and, in document and reports, the possibility of information overload.*

Additional forms of communication that address the other three senses, as touch, taste, and smell, can become either a great support to interpersonal professional communications, if everything goes well, or a disaster in the relationships, if they, or their results, are not appreciated. In fact, in general, this variety of communications, including shaking of hands, giving somebody high five, putting hand/arm on shoulder, hugging, kissing, having lunch/dinner together, sharing perfumes/odors, etc., can break down some barriers in favor of better personal relationships … or lead exactly to the opposite effect (for instance, in several important cultures, people do not like to be touched at all).

Indeed, *communication medium*, which is part of the common communication environment, *is shared between the stakeholders, and, of course, it has to be available, accessible, reliable, and … less noisy as possible. In fact, all communication media introduce noise, which can distort the message, up to the point of affecting its meaning, or of making it unintelligible*, and, then, a preventive attention to this issue is essential, too, in order to minimize possible interferences. From the noise point of view, *if communication medium consists in another stakeholder, this is the most risky situation*, since people aim to add their own value and/or contribute, then tending to distort communications. Ultimately, in general, we may distinguish three main types of noise: *physical noise*, e.g., rumors that affect talks, irrelevant/wrong messages, illegible handwriting, blurred type, fonts that are difficult to read, misspellings, and poor grammar; *physiological noise*, e.g., degradations such as loss of vision, and/or hearing; and *semantic noise*, e.g., when medium is not actually located in common communication zone, and/or coding/decoding errors are present.

Finally, to complete the overview on the stakeholder interpersonal communication model, there are some topics that are specific for either the transmitting or the receiving stakeholder. At first, *all project communication are purposeful*, and, then, *transmitting stakeholder has to convince, or to persuade, receiving stakeholder about the quality of his communication.* Even though project management literature focuses mainly on project communication as a mean to transmit (and other) project information, *several other basic purposes are possible for project communication:*

- *providing project information;*
- *obtaining project information and/or data and information that are relevant to the project;*
- *requesting an action, and/or preventing an action, and/or starting an action, and/or changing an action;*
- *providing support to decision, and/or advising, and/or consulting;*
- *learning, and/or facilitating learning;*
- *setting up and/or improving relationships.*

Whatever the purpose of the communication is, convincing the recipient stakeholder about the quality of the communication, in order to influence his following behavior, requires persuasion skills, and the ability of making use of those "*weapons of influence*" that can generate in him a sort of "*psychological positive automatic response*" (Cialdini, 2007):

- *reciprocation* ("the old give and take ... and take"). The perception of receiving, as it is a gift, something unexpected, stimulate both positive engagement and the wish to somehow reciprocate; this type of response can be initiated, for instance, by offering and/or proposing something of a perceivable value "first";
- commitment and consistency ("hobgoblins of the mind"). Defining properly both commitments and relevant consistent rewards greatly stimulates positive engagement;
- *social proofs* ("truths are us"). Presenting a community in which experts, and/or satisfied users, and/or networking, and/or celebrities are included, and stimulating the sense of belonging to it, can be a great support to stakeholder engagement and/or approval and/or action;
- *liking* ("the friendly thief"). Liking principle generates positive responses through factors as empathy, friendship, attractiveness, cooperation, etc.;
- *authority* ("directed deference").This principle is based on the fact that people aim to follow credible experts, managers, technologists, or, in general, professionals who look like they know what they are saying; and
- *scarcity* ("the rule of the few"). Things that are rare, or available with difficulty, are commonly considered better than the others are; showing an approach of "unicity" versus the receiving stakeholder may become a success factor in the communication.

Another skill of greatest importance that is specific for receiving stakeholders, but that is needed for transmitting stakeholders too, is the capability of *active listening and observation.* In general, *active listening*, which in project management literature is defined as the set of techniques including "acknowledging, clarifying and confirming, understanding, and removing barriers that adversely affect comprehension" (Project Management Institute, 2017), together with *active observation, require to be approachable, tolerant, acceptant, receptive, assertive, attentive, respective for the speaker, and for his verbal/paraverbal/nonverbal behavior, and all that, in case of face to face communication, by ensuring and maintaining a constant contact (including eye contact) too.*

Definitively, the essential issue of initiating, developing, maintaining, and improving an effective communication with key stakeholders, in order to manage successfully the relationships with them, requires a comprehensive set of interpersonal skills, which should include, without being necessarily limited to, the following:

- awareness that *key stakeholders are persons* (and not transmitting/receiving devices!), *with their own roles and expectations, requiring also a consequent capability of expressing (not only feeling!) respect for each of them and for them all*;
- capability of *expressing respect for stakeholder contents and issues* too, *also by signaling availability to the listening, receptiveness, acceptance, understanding, and by making questions and/or asking information*;
- awareness that *we are all stakeholders*, and that, at the same time, *we are not the shareholders, i.e., the owners, of the relationship*, and *consequent capability of behaving accordingly*;
- awareness that *stakeholders may be interested to our own contents, but they usually feel bothered if we report third parties' contents, as much as they may be interested to our professional self, but they usually feel bothered if we show an "image" of it* (that, anyway, could be discarded as "false" by paraverbal and nonverbal language interpretation);

- awareness that *stakeholders may have different individual and organizational frames of reference and use diverse languages, and consequent capabilities of both pulling our own frame of reference off-center in order to tune in theirs, and using their own business/organizational/ spoken (this latter if possible) language, also in terms of goals, risks, and opportunities, and in verbal, paraverbal, and nonverbal terms*;
- awareness that *each communication is an exchange of contents*, and consequent *capability of leaving always something available to other stakeholders contributes, because this will support communication effectiveness*;
- capability of *being open mind, also abandoning prejudices and mental models, and of building shared visions*;
- capability of *understanding the other stakeholder scope, of identifying, and/or avoiding, and/or removing eventual blocks and/or barriers, and of targeting/reaching harmonically common agreeable solutions*;
- awareness of *stakeholder feeling that they do never have time, and that their available time may be asynchronous with respect to other needs and vice versa, with consequent capabilities of being synthetic, by going directly to the point, being clear and concise (also by preparing executive summaries and/or dashboards, and minimizing introductions and/or prefaces), and, in general, by planning and managing the times of the relationship (without being managed by them)*;
- capability of *being analytic, distinguishing facts from opinions, and/or causes from effects, being also neutral versus emotions*;
- capability of *organizing and managing the appropriate communication environment, taking care of reception, interactivity, and privacy, minimizing noises, and setting up a balanced mix of formality and informality*;
- capability of *taking commitment to answer*, and, then, *answering coherently and appropriately*;

- capability of *being always positive* (glass is always half-full, and is never half-empty!);
- awareness that stakeholders *prefer to hear proposals that target success, rather than problems to be solved* (especially the problems that we are supposed to solve);
- *awareness that the choice of the titles determines both the attitude and the attention of stakeholders*, and consequent capability to define the titles properly;
- awareness that *stakeholders may be/are interested in a possible engagement for present and future times, while they do normally perceive communications about past times, in which they were not present, just as self-referential*; and
- *awareness that a confusing communication is very risky, and that it could be useful just to gain some time.*

As a conclusion of this chapter, I propose the following paradigm for effective communication:

"we should communicate ourselves, but in the language of others".

In this way, other people will appreciate that we are true persons, who expose contents in which we believe … and, since we are using their language, they should understand what we do mean too, so that, definitively, communication will reach its purpose.

Chapter 6

Stakeholder Network Management: Informative and Interactive Communication

Direct interpersonal communication is essential for managing closely relations with key stakeholders, but, since it is expensive in terms of both time and effort, it cannot be extended easily, effectively, and efficiently to other non-key stakeholders. In fact, we saw in previous chapters that, in each project, *the stakeholders' relations are characterized by an intrinsic multilevel complexity*, and this is also because stakeholders are numerous, and their relations influence each other. Therefore, we have *not only many relations with stakeholders to manage, but also a large variety of relations among stakeholders to monitor, and this makes quite difficult and/or impossible to use forms of direct communication with the totality of non-key stakeholders too.* The above-mentioned communication indirectness involves mainly two aspects: first is the *"plurality" of communication*, meaning that communications with non-key stakeholders are generally in the form either of one-to-many, in case we refer to the project manager communicating with all other stakeholders, or many-to-many, if we refer to communications among stakeholders; second aspect is the *communication need of being channeled and someway mediated*, which is presently and actually solved by the large domain of Computer-Mediated Communication (CMC).

Nowadays, indeed, we are all interconnected, and, of, course, this is valid for stakeholders too; *all stakeholders are networked*, and this was a sort of "silent revolution" in managing project communications, which had, and has, a considerable impact on project success, and also on project risks, i.e., on both opportunities and threats. *The above-mentioned "silent revolution" happened because the original, and still principal, goal of project communication is sending properly project information, i.e., contents, rather than developing interactive relationships with stakeholders, and this led, in project management, to forms of "cold" communications that were basically thought as "impersonal" and characterized by a very limited interaction, while the apparent paradox of the network is that, although it is a "cold" additional technological mediation, it realizes indeed the individual addressability and connectivity, then allowing a new "warmer" interpersonal interaction among people, i.e., stakeholders.* However, to go in depth about this issue, it is important to make an overview about the complex, often ambiguous, and someway difficult, relation between project communications and project management.

In general, starting from the early beginning (Project Management Institute, 1996), communications in project management were mainly focused on the broadcasting of project information, mostly in the formats of reports, and, more recently, in the formats of dashboards and/or scorecards too (Kerzner, 2015). While almost everybody agrees today that the weight of relationships is determinant for communication efficacy, although the presence of qualified contents is anyway necessary, above form of communication is evidently cold, purely informative, few interactive, and very similar to other forms of vertical and/or hierarchical, communication, which all have the problem of leaving unchanged, and/or even increasing, both the distance and the barriers between sender and receivers. Although the emergence of interactive forms of communication and, especially, the accreditation of both stakeholder as a subject group (International Organization for Standardization, 2012) and of project stakeholder management as a knowledge area (Project Management Institute, 2013) gave a strong positive impetus on relationships improvement, *the present vision of the recipient stakeholders as few interactive participants in the communication is still almost exactly the same of more than twenty years ago* (Project Management Institute, 1996 and 2017): "*The receiver is responsible for ensuring that the information is received in its entirety, interpreted correctly, and acknowledged or responded to appropriately*". The above-mentioned perspective, on the one side, hinders effectiveness of communication, which of course is, as almost all other project issues are, a responsibility of project manager, on the other, encourages at most limited feedbacks, rather than, as we saw in previous chapter, a useful exchange of information (including, e.g., obtaining project information and/or data and information that are relevant to the project, requesting an action, and/or preventing an action, and/or starting an action, and/or changing an action, providing support to decision, and/or advising, and/or consulting, learning, etc.). *Indeed, while, on the one hand, the forms of project communications that have been someway prioritized included unilateral, impersonal "broadcast" of reports; on the other hand, the Internet, by offering a common medium that supports several evolutionary forms of CMC, overwhelms traditional approaches. All stakeholders are now individually addressable, interaction is embedded, although quite often with a certain time delay, and the shareable information are powered, since they may include also project "promotional" information, and, moreover, useful lessons learned.*

Therefore, Interactive Stakeholder Communication Model evolves and spreads in a stakeholder network; in each project, communication streams are originated by each stakeholder community, flow in common media, are exchanged, stimulate other communication streams, and so on (see Figure 6.1). Indeed, communications with and among stakeholders are, of course, absolutely complex, but the segmentation of stakeholders in four communities drastically reduces the complexity in their management, because the huge domain of communications with and among stakeholders, can be represented by six basic communication streams, which are, respectively, suppliers–purchasers, suppliers–investors, suppliers–influencers, purchasers–investors, purchasers–influencers, and investors–influencers, and which relate each other via an actual network, i.e., the stakeholder network.

It is important to be aware that stakeholder network (Pirozzi, 2017) has specific characteristics, which have to be managed carefully; indeed, the network is not neutral with respect to the project, and it may turn out to be either the originator, or the amplifier, of both great opportunities and critical threats. In fact, stakeholder network:

- *is a fast multiplier of the stakeholder satisfaction and dissatisfaction messages, then generating a snowball effect about their feelings versus the project*; marketing literature generally reports that a satisfied stakeholder communicates it on average to three others, while a dissatisfied stakeholder communicates it on average to other ten, and these phenomena on the network are substantially immediate;
- *is multilingual*, since it carries messages in different languages, including project management language, business languages, economic languages, media language, natural languages, spoken languages, social media languages,

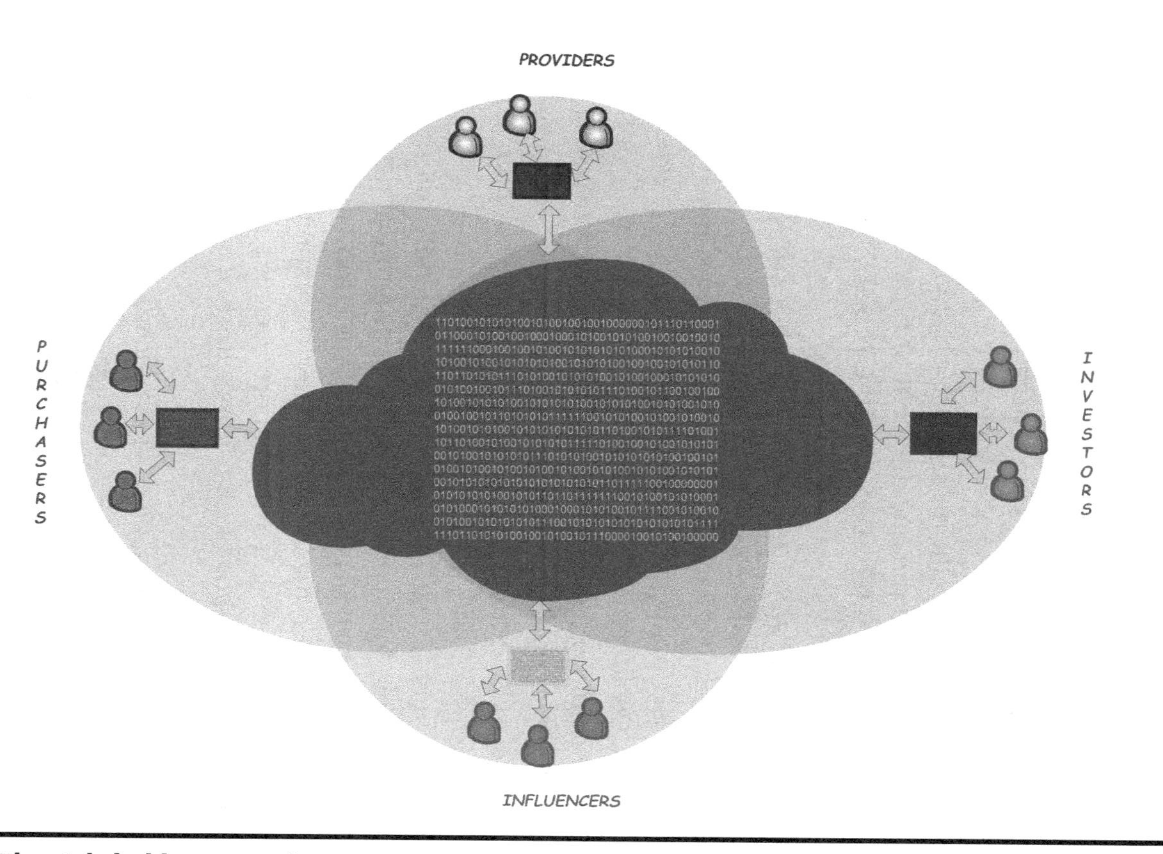

Figure 6.1 The stakeholder network.

paralanguages, etc., *and it allows a large variety of multiple and/or integrated, and/or hybrid formats too*;

- *although its life cycle basically coincides with the project life cycle, since it was born with the project and it evolves over time with it, a part of the network was evidently already alive in previous investment life cycle, and another part will survive even after the project completion, in project/product/service life cycle*; the above-mentioned "long life" has significant and/or potentially critical impacts on both the management of project stakeholder relations and on the success of the project itself, as we will see in the following chapters;
- *is both informative and interactive, and has a 2.0 behavior through and through*;
- *is not controlled by any project stakeholder* (much less by the project manager!), *but it is continuously influenced by all stakeholders*; *stakeholder network is extremely noisy, since communications among stakeholders seriously affect each other, and monitoring relations among stakeholders becomes essential to manage effectively relations with stakeholders*; and
- *is asymmetric*, e.g., while providers generally transmit reports that include somehow large amount of data, and that are characterized by a consistent information richness, they often receive from other communities of stakeholders relatively few information, even if these are of high quality and/or of high importance.

In stakeholder network, CMCs, which include a large variety of messages, e.g., e-mails, text/visual/video/audio/hybrid social media messages, instant messages, video/audio/text chats, websites, online forums, blogs, wikis, phone calls, etc., *are numerically the major part of communications. Nowadays CMCs became, in most cases, either essential itself or an essential support for communications with, and among, the four communities of stakeholders, i.e., the Providers, the Purchasers, the Investors, and the Influencers.* For better or worse, CMCs are not only "transient, multimodal, with few codes of conduct governing use, and allowing for a high degree of end-user manipulation of content" (McQuail, 2010), but these are also fast, irreversible, recordable, extremely numerous, and often evolving and/or mutating, both in their formats and in the language/paralanguage that they use.

In case of communication with key stakeholders, who generally include project team, customers, project sponsor, and top management, *while interpersonal direct communication, especially in the form of face to face communication*, as we saw in previous chapter, *due to both its information richness and its orientation toward the development of the relationships, can be considered at top of effectiveness, CMCs* (with the exception of video chats/conferences, which can be considered as a "lower quality" face-to face communication), *are in any case of extraordinary importance, to support initiating, developing, and maintaining interpersonal communications. CMCs are, in fact, minimally invasive, and their use both decreases the possibility of making fatal communication mistakes and increases the possibility of succeeding in the organization of appointments, meetings, shareable agendas, and so on.* Specifically, *CMCs are a powerful help both to initiate interpersonal communications, since today a greatest communication risk is that in cases a previous interpersonal relationship does not exist, stakeholders tend to let not opening a communicative relation, and to maintain communications, since their use can be quite continuous, but still cheap and a little invasive.* On the other side, we see an *overuse of CMC, and specifically of e-mails, in communications with key stakeholders; the easiness of a quick addressing, due in particular to the button "answer to all", can often lead to a correspondent easiness of altering the appropriate sequence of recipients, then generating the risk of inefficiently confusing "for action" and "for information" stakeholders addressees.* Furthermore, *due to the facts that both incoming e-mails, if they do not answer to a specific request, are generally associated with routine information, and that e-mails tend to generate in our daily personal and professional lives spamming and/or information overload, there is the greatest risk that their content is not considered as important, or not processed at all. In addition, CMCs necessarily require the use of a synthetic written language, which unavoidably filters all those paraverbal and nonverbal elements that are essential in effective communication with key stakeholders, often introduces semantic noise, and even increases enormously the risk of fatal mistakes and/or misunderstandings in those phases in which careful negotiating and/or managing of conflicts are needed.* Definitively, *in case of communicating with key stakeholders, CMCs may be a powerful support to interpersonal direct communications, but they can never replace them, especially if there is the need of exchanging important information.*

Anyway, in case of communications with non-key stakeholders, stakeholder network and CMC are a great opportunity of usability and efficacy, and moreover, in case of communications with the community of Influencers, they are almost always the unique possible way to reach them. Project communications spread to a larger domain of stakeholders, and this may be a further opportunity to support the project success by both promoting and defending the reputation of project itself. Indeed, *in last years, the phenomenon of the trend of a continuously increasing importance, especially in large and major projects, of those stakeholders who are not directly contracting parties, but who are affected and/or feel to be affected by the project, led to the inclusion, in the domain of those stakeholders whose satisfaction is a critical success factor for the project, of the behavioral community of Influencers, too. In fact, the diverse worlds* of culture, communication, instruction, politics, public health, non-profit organizations, including local communities, the web communities, the associations, the trade unions, the media, the authorities, the central and local public administration, the regulatory bodies, the potential customers and users, the participants and the candidates to participate in the project, *can very rapidly become either great supporters, or relentless opponents, of the project, and moreover, each of the above-mentioned influencers can switch from one role to its opposite as quickly.*

Therefore, especially in major, large, and complex projects, after the ever-present traditional reporting, and the finally-recognized-as-basic interpersonal direct communication with key stakeholders, a third main type of stakeholder communication emerged decisively: *the project relations*, or project marketing (Bourne, 2015). *Project relations involve both informative and interactive communications, and the main purposes of this type of communications are*

- *adding value to the project, by positively informing stakeholder network, mainly about project's scope, core and boundaries of which may be unclear, and about project progress, and, moreover, by promoting project's reputation, which became one of the major "competing constraints" (Kerzner, 2015), as much as the brand's reputation of the organization that realizes the project*;
- *preventing, to the greatest extend possible, misinformation, false expectations, and all kinds of rumors, by using the power of effective information*; and
- *defending, in almost real-time, the reputation of the project, the project manager, the project team, and the organization, from every type of attack and/or improper influence, and satisfying additional or deeper stakeholder information needs that may occur, by monitoring continuously stakeholder network, and by interacting with it as soon as possible from the perception of an event.*

A further extraordinary potential usefulness of stakeholder network is the capability of enhancing the lessons learned which are available in each organization *with those contents that are shareable on web-hosted knowledge bases.* Great global independent websites as *PM World Journal* https://pmworldjournal.com/ and

PM World Library https://pmworldlibrary.net/, as much as the websites of both International and National Project Management Associations, have a very important role, and may have an even major one, in spreading and enhancing both the discipline and the culture of project management, and they should be encouraged as much as possible.

Definitively, since the stakeholder network is not neutral with respect to the project, and it may turn out to be either the originator or the amplifier of both great opportunities and critical threats, project manager cannot be neutral, and cannot act neutrally, versus the stakeholder network itself, which has to be respected and properly addressed. In general, in major and/or large and/or complex projects, while the design, the organization, the development, and the improvement of an appropriate project relations campaign may require specific skills, project managers must have at least those 1.0 and 2.0 competencies that allow continuous and effective monitoring and interaction with stakeholder network. Ultimately, project manager should

- *deal with the stakeholder network as it is a project in the project*, by initiating, planning, implementing, controlling, and closing all the necessary communicative actions, *in order not only to be influenced by the stakeholder network, but also to influence it*;
- *communicate effectively also by using with each stakeholder a common shared language*, so learning, using as appropriate, and improving the basic languages that are needed beyond project management language, including the languages of customers and/or users business, of general management, of economics, of worldwide web, of social/and or other media;
- *manage appropriately the relationships with both key stakeholders, by using interpersonal direct communication as primary, and non-key stakeholders, by using CMC as primary*;

- *take into account that stakeholder network is alive, evolutive, that it was born in the investment life cycle before the project was, and that it will survive also in the product/service life cycle after project completion*;
- *develop and/or maintain competencies in terms of proactive monitoring of stakeholder network, including the essential capacities of detecting the signals of satisfaction and/or dissatisfaction, of amplifying weak signals and/or of performing checks of satisfaction/dissatisfaction, of defending reputation, of answering in almost real time, of conveying appropriate project information, of defining/measuring/sharing with stakeholders appropriate key performance indicators (KPIs), of managing relationships with hostile and/or negative stakeholders too, by communicating those information that tend to cause their disengagement, and their dissatisfaction*; and
- *manage effectively the quality of information, by supporting the project with appropriate and well-presented information and by combating fake and/or negative information, the quantity of information, by preparing appropriate executing summaries, dashboards, scorecards, and by avoiding phenomena of information overload, and the communicative mix, by integrating appropriately informative, interactive, interpersonal, direct, computer mediated, face-to-face, remote, verbal, non-verbal, vocal, written, audio, visual, hybrid communications.*

Managing relationships with stakeholders is evidently very complex, and, then, every project manager should necessarily add to high competencies in effective communication, which are foundational for initiating, developing, and managing effective communications with the various stakeholders, those basic personal and interpersonal skills in terms of personal mastery, leadership, and teaming, which are necessary for his/her optimal, both personal/professional and organizational, behavior.

Chapter 7

Basic Personal and Interpersonal Skills: Personal Mastery, Leadership, Teaming

Managing effectively both relationships and communications with stakeholders who have so diverse interests, expectations, and characteristics requires modern and advanced interpersonal skills; however, *the development of interpersonal skills is not only a consequence of learning of tools, methodologies, and techniques, but it relies on a continuous individual learning path that is based on increasing personal knowledge, abilities, and experience, and, then, the ownership and the continuous improvement of personal skills are needed, too. Personal skills and interpersonal behavior integrate in a "professional self", whose basic continuous-learning approach to stakeholders has to be the result of the growth of both "how to be" and "how to behave" skills*; therefore, *personal mastery is essential* to improve organizational behavior too, and it becomes the foundation also for developing those leadership and teamwork that are crucial elements of project management starting from its early beginning up to today.

Personal mastery is one of those four core disciplines (the others are mental models, shared vision, and team learning) of learning organizations that are integrated in the *fifth discipline "System Thinking"*, and it can be defined as *"the discipline of continually clarifying and deepening our personal vision, of focusing our energies, of developing patience, and of seeing reality objectively"* (Senge, 2006). Although personal mastery is based on both the development of competencies and/or skills, and on spiritual growth, it goes beyond all of them (Senge, 2006) in the direction of *proficiency*; indeed, it integrates two perspectives that are foundational in managing stakeholder relationships, communication, and project itself, which are, first, *the focus on objectives to be achieved, i.e., answering to the question "what is really important", and, second, the continuous learning approach based on the clear and updated assessment of the reality, i.e., answering to the question "what is real". Actually, in stakeholder relationship management, the assessment of the real situation is fundamental in stakeholder identification and analysis processes, while focus on objectives to be achieved and to expectations to be satisfied is essential in stakeholder management processes.*

Moreover, personal mastery is the basic state to enable that *self-awareness, which is not only one of the four domains of emotional intelligence but also the foundation of the other three, i.e., personal management, social-awareness, and relationship management.* The above-mentioned emotional intelligence groups of competencies are crucial for leadership, and specifically for that *resonant leadership*, which *represents the great and effective ability of connecting personally with the followers* (Goleman, Boyatzis, and McKee, 2002) *and/or, in project management, with the stakeholders.* In fact, each one of the four emotional intelligence domains interacts with others (see Figure 7.1) and includes a set of core leadership competencies, as follows.

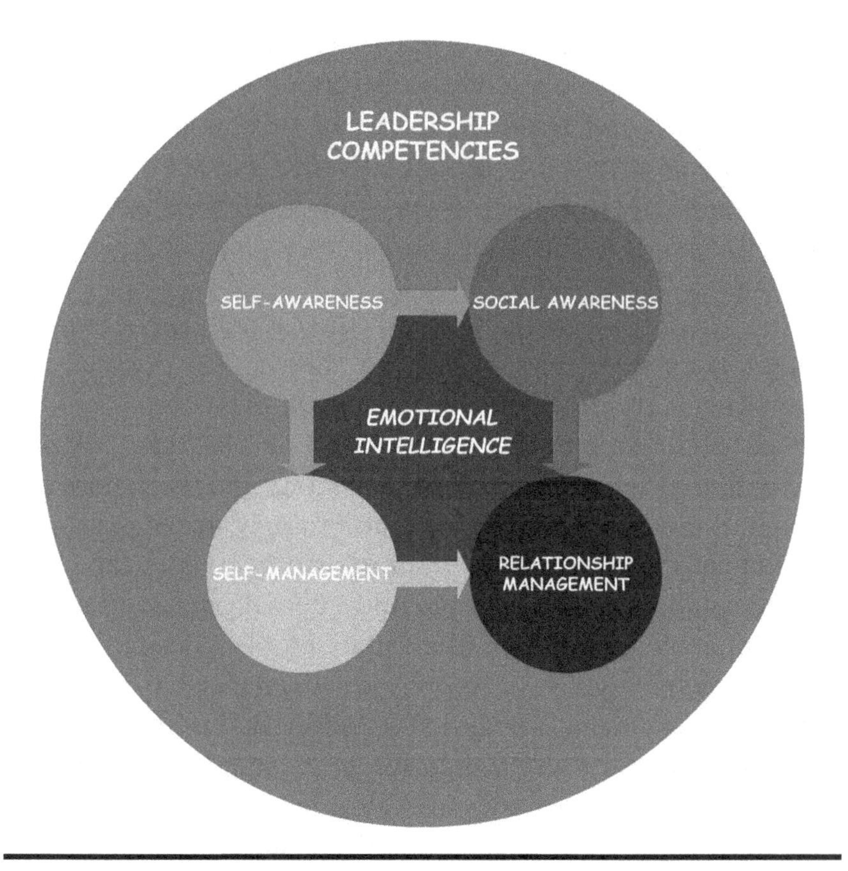

Figure 7.1 The emotional intelligence leadership competencies.

The personal competencies refer to the capabilities of managing ourselves, and they consist in *self-awareness* and *self-management* competence groups. The *self-awareness* competencies group includes the *emotional self-awareness*, which is the ability of understanding own emotions, of estimating their possible impacts, and of using "gut sense" as a decision-making support, the *accurate self-assessment*, which is the ability of understanding personal strengths and weaknesses, and the *self-confidence*, a sound sense of personal value and capabilities. The *self-management* competencies group includes *emotional self-control* (the ability of keeping under control all kinds of emotion, especially the disruptive ones), *transparency* (the ability of demonstrating ethics and trustworthiness), *adaptability* (the ability of being flexible in order to adapt to different situations and to overcome obstacles), *achievement* (the capacity of continuous personal improvement toward excellence), *initiative* (the ability of reacting properly and quickly to events and of being proactive too), and *optimism* (the ability of getting the upside in every situation).

The *social competencies* refer to the capabilities of managing relationships with others, and they consist in *social awareness* and *relationship management* competence groups. *Social awareness* competence group includes *empathy* (the ability of understanding the emotions and the perspectives of the others, and of acting accordingly in the direction of their satisfaction), *organizational awareness* (the capability of understanding both external and internal organizational environments) and *service* (the capability of understanding and meeting follower, client, or customer needs [and, in project management domain, stakeholder needs and expectations]). The *relationship management* competence group includes competencies and skills that can be all directly related to project management, and that all are of extraordinary importance in it, such as *change catalyst, developing others* (coaching), and, above all, *conflict management, inspirational leadership, teamwork and collaboration, and influence.*

Influence is a foundational concept of *leadership*: "*Leadership is influencing people to take action.* In the workplace, leadership is the art of getting work done through other people. Leadership can be widely distributed within an organization—most everyone leads at some time or other, if not all the time. And it's highly situational: anyone might step forward to lead, given the right circumstances" (Goleman, 2012). Leadership and influencing are key points in project management literature starting from its early beginning. In the first edition of *PMBOK Guide* (Project Management Institute, 1996), leading involves establishing direction, aligning people, motivating and inspiring, while influencing the organization involve the capability of "get things done". Presently, in the sixth edition of *PMBOK Guide* (Project Management Institute, 2017), "leadership skills involve the ability to guide, motivate, and direct a team", and *Leadership is*, together with Strategic and Business Management, and Technical Project Management, *one of the sides of the PMI Talent Triangle*™, which is considered by Project Management Institute the ideal skill set—*while influencing is specifically one of the major leadership competencies*, together with brainstorming, coaching and mentoring, conflict management, interpersonal skills, listening, negotiation, problem solving, team building, and, conveniently, emotional intelligence.

In this book, we consider that *leadership competencies are effectively applicable not only to the traditional domain of those specific key stakeholders, who form the project team, but also, in a much greater sense, to the whole domain of project stakeholders.* Indeed, *leadership skills are not only put into practice by the project manager and/or, if this is the case, by other stakeholders, but they are also perceived, and quite immediately recognized, by all stakeholders.* This usually happens mainly either via verbal, paraverbal, and nonverbal language, whether an interpersonal direct relationship exists, or via records, writings, information about reputation and/or satisfaction of expectations, in case of the other non-key part of stakeholder network is involved. Definitively, *leadership is basic to generate trust, and, then, to manage effectively relations with all types of stakeholders.*

Practicing leadership in project management requires the flexibility that is necessary to adapt to the different stakeholders and/or situations, so that *leadership repertoire of project manager should include the six leadership styles: visionary, coaching, affiliative,* and *democratic,* which are the "resonant" ones, *plus pacesetting* and *commanding*, which can be very useful and/or necessary in some specific situation, but which often have negative impact on the working environment (Goleman, Boyatzis, and McKee, 2002). In more detail:

- *The visionary style is best when a new, or a clearer, direction is needed*: it motivates people versus achieving common goals, by using competences of empathy, self-esteem, sharing of information, and the capability of catalyzing changes, and it has a strongly positive impact on climate. It may be less effective in case of highly experienced people, who can consider it redundant, and unnecessary.
- *The coaching style is best to help people to enhance their medium-term performances via building their long-term capabilities, connecting personal and organizational goals, and sharing strengths and weaknesses*: it uses mainly the competencies of self-awareness, empathy, and human resource management, and it has a highly positive impact on climate. It may be less effective in case of people who are not so much motivated in their professional growth, and, since it has medium-term positive effects, when time constraints are strict; moreover, there is the risk that if this style is not realized properly, it may be perceived by people as micro-management, and then, it may be refused by them.
- *The affiliative style is best to get through stressful situations, and/or to strengthen relations, and/or to heal eventual rifts*: it creates, and increases, harmony by using mainly the competencies of empathy and effective relationships management, and by connecting people to each other, and it has a positive impact on climate. It may be less effective when there are strict objectives in terms of quality to be incorporated, and/or of people growth, and/or of mistakes to be corrected.
- *The democratic style is best to obtain inputs and/or approvals from people*: it values the inputs of the people and it obtains their commitment and contributions

through the participation, by using mainly the competencies of effective communication and of enhancing cooperation, and it has a positive impact on climate. It may not be so much effective in case it leads to endless meetings and/or to a certain degree of indecision.

- *The pacesetting style is best to achieve results from an already motivated and competent team*: it builds and meets for people exciting goals and/or challenges, which generally represent excellence, by using mainly the competencies of effective guidance toward objectives to be achieved, of initiative, and of thoroughness. Since it is a lot demanding from everybody, it can depress less competent and/or motivated people that suffer comparisons, and since it is so often poorly managed, its impact on climate may be frequently highly negative, and it has to be used carefully.
- *The commanding style is best when there are either start-up, or critical, or involving big changes situations, and/or when there are unresponsive and/or problematic people involved, and/or when time to act and/or react is very limited*: it gives clear directives, and it demands that they are promptly and duly satisfied, moreover it requires the acknowledgment of the power and/or the role, and it uses mainly the competencies of self-control, of initiative, and of effective guidance toward objectives to be achieved. Since it tends to inhibit flexibility, and to decrease the people's motivation, and it is so often misused, its impact is quite frequently highly negative, so that this style has to be used carefully too.

Definitively, leadership is a primary issue in diverse project management perspectives:

- "Project manager is both a leader and a manager of project activities" (The International Organization for Standardization, 2012).

- "Leadership skills involve the ability to guide, motivate, and direct a team. These skills may include demonstrating essential capabilities such as negotiation, resilience, communication, problem solving, critical thinking, and interpersonal skills" (Project Management Institute, 2017).
- Leadership is a primary "competence element" of the "competence Area" people, and it "means providing direction and guidance to individuals and groups. It involves the ability to choose and apply appropriate styles of management in different situations" (International Project Management Association, 2015).
- "Project management effectiveness depends upon the project manager's authority, capability, knowledge, and leadership skills" (Archibald and Archibald, 2016).
- "Today, with Project Management 2.0, project managers are being asked to function as managers of organizational change on selected projects. This approach is now being called transformational project management leadership" (Kerzner, 2015).
- "The trend is moving toward widespread project management where the project leadership is shared among the entire project team" (Sampietro and Villa, 2014). *Repositioning leadership and followership* (Dalcher, 2018) becomes very important: "project followership means proactive participation in all managerial aspects of the project work within an individual's visibility horizon" (Sampietro and Villa, 2014).

The project team members are actually the key stakeholders who are essential to have things properly done, and to address, and then to achieve, both project objectives and other stakeholders' satisfaction. Therefore, *the importance of the team in the projects is nowadays continuously increasing, and managing properly the team dynamics became essential.* In fact, a team is not only composed by living beings, but it is a living entity itself, with its own specific life cycle, and its own

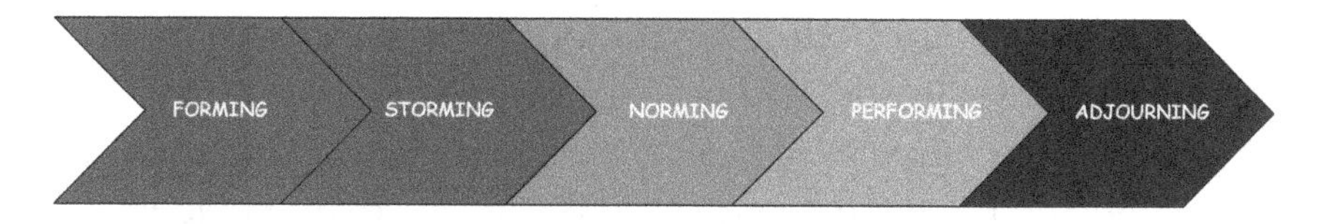

Figure 7.2 The basic stages of every team life cycle.

organizational behavior. In each project, as in each organization, the results of a team should be major than the as-is sum of the results of all its member, and, then, a positive and effective integration of all the contributions to be made by the project manager becomes a success factor; in fact, team building encourages the team members to cooperate effectively, and it requires a careful attention to team's life cycle.

In order to facilitate teambuilding, it is essential to be aware that, in every team life cycle, there are five basic stages (Tuckman, 1965), each one of which is peculiar with respect to other (see Figure 7.2):

- forming;
- storming;
- norming;
- performing; and
- adjourning.

Basically, project manager should support adequately the team in all of its life cycle in order to have the team performing effectively as soon as possible, since, in any case, this condition must start before the completion of planning processes; in order to do this, taking into account the specific characteristics of each stage, project manager uses several different leadership styles, as appropriate. In the stage of *forming*, team members start to know each other with respect to the project, and to put their different emotions, mental models, and visions in the project itself, with a consequent risk of generating divergent paths. In this stage, a leadership commanding style is appropriate to get that initial trust that is needed to proceed, and to start to share both a unique project vision and common project objectives to be achieved. In the stage of *storming*, team member compete, sometimes intensively, to gain each one a specific role in the project that can be acknowledged by all the others, and, if they do not satisfy their expectations, they resist to project progress. In this case, leadership style of coaching is most appropriate to get out from this phase as soon as possible, and, anyway, before the completion of planning group of processes. In the stage of *norming*, people build their shared visions, specify common behaviors and "rules of engagement", and a visionary leadership style may be most appropriate to facilitate the process. In the stage of *performing*, which is finally the one that characterizes the effective cooperation, an affiliative, delegating, leadership style may be most effective. This style may probably be the best also in the *adjourning* stage, which is the one that will lead to the dissolution of the team. Before reaching the adjourning stage, it is extremely important that project manager obtains from all other managers involved the proper assurances about the "day after" the project completion and the professional future of all the team members; otherwise, there will be a high risk of project consistent delays.

In project management, *teamwork is the most evident demonstration of stakeholders' power, since effective cooperation may greatly improve productivity up to levels that are still partially unknown.* In the projects, the process of *integration*, which is basically the same that leads to the organizations' birth, is quite well known: *the results of the teams may be, and so are expected, greater than the sum of the individual contributions*, due to their integration, which is mainly a project manager's task. In addition, *project management tends to increase also the individual productivity* by enhancing the concept of *responsibility. Team members are transparently charged with the responsibility of work packages, and this happens not necessarily because of their belonging to an organizational structure, but on the basis of a shared decision process that is based on both individual competence and common effectiveness recognition, and that is also made known and official to the other stakeholders.* The above-mentioned process guarantees an improvement of both self and social awareness and, therefore, it enhances motivation, and it leads to better both individual and collective results; moreover, maybe it is not a coincidence that the *community of project managers identifies, in its perspective, the responsibility as a major ethical topic* (Project Management Institute, 2006).

Chapter 8

Ethics in Stakeholder Relations

Ethics are specifically integrated in each project, and are essential both in project management and for project managers; indeed, codes of ethics and of professional conduct always characterized project management and project managers' communities, so that their guidelines, and also their peculiar nuances, can address both the importance and the effectiveness of ethical issues in stakeholder relations.

Main values that support Project Management Institute (PMI's) Code of Ethics and of Professional Conduct (Project Management Institute, 2006) are *responsibility, respect, fairness, and honesty. Responsibility* is defined as "our duty to take ownership for the decisions we make or fail to make, the actions we take or fail to take, and the consequences that result"; its mandatory standards include both being informed and informing about, upholding the applicable laws, regulations, policies, and rules, and reporting eventual illegal or unethical conduct. In addition, responsibility's aspirational standards include decision-making and action-taking, which are based on the best interest of "*wider stakeholders*" as the environment, the public health, and the society in all respects, and include also the acceptance of only those assignments that are relevant to really owned qualifications, competence, and skills. *Respect* is defined as "our duty to show a high regard for ourselves, others, and the resources entrusted to us. Resources entrusted to us may include people, money, reputation, the safety of others, and natural or environmental resources"; its mandatory standards include negotiations in good faith, non-abusive actions versus others, refraining of exercising inappropriate decisions and/or actions that are directed toward personal benefits, and the respect of all kinds of property rights. In addition, respect's aspirational standards include information about others' norms and customs and consequent appropriate behavior, listening to others and trying to understand them, direct approaching of people there is a conflict and/or misunderstanding with, and professional conduct—in all cases, i.e., *versus negative stakeholders too*. *Fairness* is defined as "our duty to make decisions and act impartially and objectively. Our conduct must be free from competing self-interest, prejudice, and favoritism"; its mandatory standards include both information and management, as appropriate, of existing and/or potential conflict of interest situations, while fairness' aspirational standards include transparency, impartiality, objectivity, and behavioral equity. Finally, *honesty* is defined as "our duty to understand the truth and act in a truthful manner both in our communications and in our conduct"; its mandatory standards involve not engaging in dishonest behaviors intending to get a personal gain at others' expenses, and, in addition, condoning and/or not engaging behaviors that are thought to damage others. In addition, honesty's aspirational standards include earnestly seeking to find out the truth, truthfulness in conduct and/or communication, creation of relationships that not only encourage truth, but also timely and accurate information, and good faith.

IPMA's Code of Ethics and Professional Conduct (International Project Management Association, 2015) addresses the core values such as *integrity, accountability,* and *transparency.* It may be interesting to notice that, according to the Code, *the commitment to acting ethically* both *leads to better projects, programs, and portfolios, and serves to proceed with the profession, and to promote it.* In other words, *while it is evident that ethics are based on values, the principle that has also to be clearly stated is that ethics add value to the projects.* Moreover, the project management community has to welcome the variety and the sensitiveness of different political, cultural, and moral contexts and/or challenges, and it has also to acknowledge that stakeholder relationships "depend upon trust, mutual respect and the appreciation of our diversity". Specific commitments to "*Project Owners and Stakeholders*" include the respect of confidentiality, the attention to others' interests, staying on guard against any biases and unethical influences, taking precautions to self-protection, the timely communication, the encouragement to critically reflect on their expectations and the ethical implications of the project outcome. In addition, specific commitments to "*Co-workers and Employees*" include the attention to the fairness of recruitments, to the highest level of health and safety measures, to reasonable privacy and personal hygiene facilities, to cultural sensitiveness, to the right salaries, to freedom about associating, and the refusal of any form of eventual unsustainable overwork and/or harmful working conditions, of forced or bonded labor, of mental or physical punishment, including any kind of harassment or bullying, of sexual harassment, and of discrimination on the basis of gender, ethnicity, religion, sexual orientation, age or on any other arbitrary grounds. Finally, it should be noticed that *ethics are addressed also with reference to the responsibility toward the wider society, the sustainability and the attention to natural environment, and the awareness of consequent educational mission.*

In the United Kingdom, APM Code of Professional Conduct (Association for Project management, 2019) reports both *Standards of Professional Conduct* and *Standards for Ethical Conduct*, and both *personal responsibilities* and *responsibility to the profession and to the Association. Standards of Professional Conduct* includes the practice of own competence in accordance with professional standards and qualifications, an honest, integer and probe behavior in relation with other professionals and/or non-professionals and/or the wider public, the attention to safety, public health, and the environment, the knowledge about relevant legislation, regulations and standards, and the relevant compliance with the consequent requirements. *The professional ethical behavior is considered in accordance with Standards for Ethical Conduct if it includes the capability of "doing things right", and in compliance with the norms of ethical behavior and public interest.* Personal responsibilities include acting honestly, respecting confidentiality, acting in the best interests of employees and clients, taking into account the wider public interests, ensuring that professional skills are kept up-to-date, claiming expertise only in those areas where it is actually present, properly informing about eventual conflicts of interests and, in case, managing them correctly, not giving and or accepting valuable gifts or undue payments, being accurate in reporting, and realistic in forecasting, being responsible, and acting with due skill, care and diligence. Responsibility to the profession and to the Association includes self-conduction to uphold and enhance the standing and reputation of the profession, behaviors that may enhance the reputation and credibility of self and/or the employer and/or the profession, non-existence of any kind of discrimination, and encouraging and/or assisting the professional development of staff and colleagues.

In Australia, the AIPM's Code of Ethics & Professional Conduct (Australian Institute of Project Management, 2018) focuses on ethical principles of *acting with integrity*, by being honest and trustworthy, by demonstrating respect for others, and by acting with a clear conscience, of *practicing competently*, by maintaining and/or developing knowledge and skills, and by acting on the basis of adequate competency, of *demonstrating leadership*, by upholding the reputation of the profession, and of *acting with responsibility*, by engaging responsibly with the community, by fostering health, safety, and well-being, and by balancing needs of the present with the needs of the future.

In Italy, ISIPM's Deontological Code (Istituto Italiano di Project Management, 2019) focuses on nine main ethical issues, which include *the respect of the laws and of the constitutional principles of the Italian State, of the European Commission regulations, of the ethics, and, in the case in which activities are developed in foreign countries, also in the respect of regulation in the country of destination*; *the commitment to respect, and to promote the respect of, the Code, with the purpose of taking care of both the Association and the professional activity*; *the update and the upgrade of professional competencies, also in order to satisfy stakeholder needs*; *the development of stakeholder relations that has to be respectful, loyal, transparent, and fair as much as possible*; *the acknowledgment and the respect of both intellectual and industrial property*; *the development of stakeholder relations that have to be based on trust, and on the respect of confidentiality*; *accepting and honoring both responsibilities and commitments*; *giving adequate information to interested parties* about eventual conflicts of interests; *respecting regulations, and feeling committed for preservation, relevant to safety, health, environment, ecological balance, and cultural, historical, and landscape heritage.*

Definitively, all ethics are relevant to stakeholder relations, and, although they introduce constraints in the projects, at the same time, since they act also as a lever that moves stakeholder satisfaction, they empower project, programme, and portfolio results. In particular, *ten guiding ethical principles* can be synthetized and suggested for application (Pells, 2015): *care for stakeholders as people, minimize harm to the planet, be honest—speak and act honestly at all times, commit to professional standards of behavior, obey the laws, avoid conflicts of interest, assume responsibility, do a good job, commit to continuous learning, growth and maturity, and, last but not least, create value for stakeholders. Indeed, ethics, and stakeholder relations, have an indissoluble link with project value.*

THE RELATIONSHIP MANAGEMENT PROJECT

Chapter 9

Stakeholder Relations and Delivered Value: An Indissoluble Link

The indissoluble link between stakeholder relations and value actually exists because all types of value are both enabled and generated by the relations among stakeholders. Indeed, starting from the early beginning:

- investment value is agreed before the start of the project, it develops during project life cycle, and it will generally cease as a capital investment at project completion, to continue as an operation and management (O&M) investment in product/service/infrastructure life cycle;
- project value develops during project life cycle, to be delivered, and then exchanged, at project completion; and
- business and/or social value, on turn, will be originated, developed, and then be exchanged in product and/or service and/or infrastructure life cycle, i.e., when project has already been completed.

In fact, *our approach to projects should always take into due consideration that stakeholder relations are, primarily, generators of value to be exchanged among stakeholders themselves, and not only an issue with an inherent complexity to be solved via stakeholder relations management.* We can summarize above-mentioned concept with the phrase "*no stakeholder relations, no value*". That is why stakeholder perspective is essential; *all types of values that are anyway generated, and which flow through the project, just like the project results themselves, are nothing but the results of relations among stakeholders, who integrate available material and immaterial resources to release consistent deliverables.*

Although, unfortunately, the attention in project management, quite often, is more or less focused on project value as if it is purely generated by project team, with a linearized approach that starts from "objective" project requirements and constraints to lead to "objective" deliverables—maybe it is not a coincidence, as we will see in depth in the following chapters, that almost 30% of the projects do not meet the initial business intents they were originated for (Project Management Institute, 2018), actual situation is different. *The process of value generation is not linear, but complex,* as everything is in nature, *and it involves, either directly or indirectly, all project stakeholders, who influence both value creation and value exchange by interacting via their relations* (see Figure 9.1).

In fact, starting from strategies phase, different stakeholder communities are present at the same time, although they may not initially interact. Investors plan investments on projects that could be profitable, customers plan to invest in projects that could support their own business, influencers monitor the situation in accordance with their own mission, and, in general, *every stakeholder community plans to develop value in accordance with its own strategy.* If different investors' and customers' expectations meet each other in a mutually convenient way, a contract or similar is signed, and a new project starts to live; *although there is for sure*, at least in this phase,

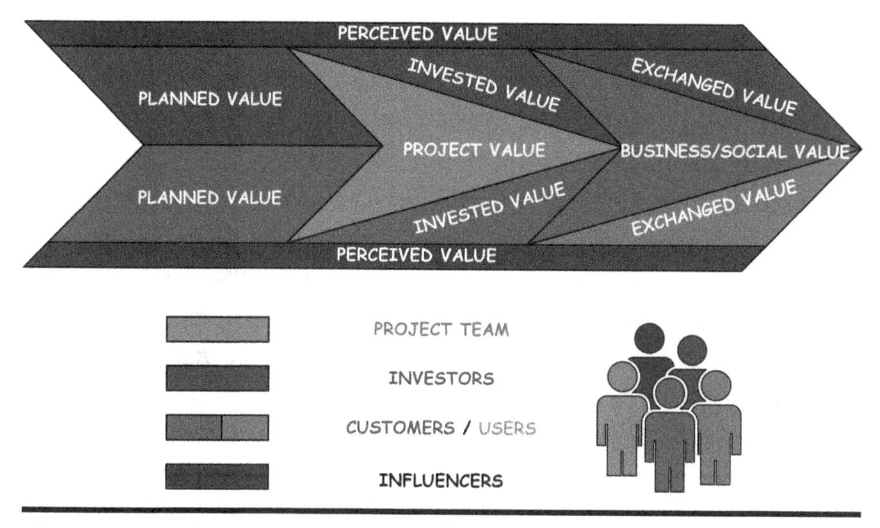

Figure 9.1 A value chain of project investment life cycle.

an agreement about project scope, requirements, cost, time, and objectives, each community of stakeholders maintains rigidly its individuality. Indeed, while *for the investors the project is the mean that is supposed to satisfy their expectations in terms of profitability and/or of social services to be delivered, for the customers the project is supposed to satisfy their expectations in terms of support to their core business,* which will be operating just in the product and/or service and/or infrastructure life cycle that will follow the project life cycle.

In project life cycle, Investors and Customers communities invest value in the form of resources, each one for its part, while Providers community, and especially project team, transforms these resources in project value, and Influencers community gives its contribution to the picture on the basis of its own *perceived value, by adding constraints and/or trying to influence the project value.* In this way, there is a continuous flow and exchange of value up to the moment in which project is completed, and delivered value is definitively released, and transferred, to customer community.

At this point, and from now on, *in the product/service/infrastructure life cycle, the customer community,* while continuing to invest in the form of production and/or delivery and/or operations and maintenance resources, *becomes the central actor for delivering value to the users in the form of products and/or services.* At the same time, *Investors community is supposed to receive value back from its investments,* while *users community exchange value in the form of resources versus products and/or services,* and *Influencers community gives its contribution to the picture, in this phase too, still on the basis of its own perceived value, by adding constraints and/or trying to influence the product and/or service value.* In other words, *when passing from project life cycle to product/service/infrastructure life cycle, there is* not only, as it was in project life cycle, an exchange of value, but, through and through, *a transfer of value* too. *This transfer of value does not imply a transfer of property only, but it causes a transfer of role, since the community of stakeholders that was a customer community in project life cycle becomes a doers community in product/service/infrastructure life cycle,* then starting to delivery products and/or services to its own users, and *it is only beginning from this phase that we can measure effectively if original project goals and business intents are actually realized.*

Indeed, the indissoluble link between stakeholder relations and value arises from different perspectives of observation, which can be synthetized in the following three principles:

- *stakeholder relations are themselves a value;*
- *stakeholder relations generate value;*
- *stakeholder relations are the mean to exchange value.*

In fact, *each contract or similar that is preliminary to each project, is actually the result of a positive negotiation among diverse communities of stakeholders, and it harmonizes their different expectations in mutually agreed scope, time, cost, and quality; accordingly, stakeholder relations generate a reciprocated commitment in terms of value to be invested, which constitutes the "initial capital" of the project, and, then, they are themselves a value, which existence and definition are necessarily prior to the correspondent project's ones.* Therefore, first principle may be summarized in *"no stakeholder relations, no projects"*. It is generally more evident that, *in project life cycle, stakeholder relations generate a value, which is incorporated in the project itself*; maybe it is less evident that generated value is the outcome of a flow of both material and immaterial value, which continuously interact and influence each other, as we will see in the following. Therefore, second principle may be summarized in *"no stakeholder relations, no generated value"*. Finally, for sure, *stakeholder relations are foundational and unique, to make concrete the exchange of value among the diverse stakeholder communities*; therefore, third principle may be summarized in *"no stakeholder relations, no delivered value"*. Accordingly, *definitively, if we focus on the project life cycle, the results of each project are nothing but the results of the relations among its stakeholders.*

Indeed, *in each project, initial value evolves toward delivered value through a non-linear path, in which diverse, but contiguous, value domains, interact and integrate, in both their positive and negative components, via stakeholder relations; these value domains include both tangible and intangible components, which are, on turn, both visible and hidden, and which, then, can be represented with an "Iceberg Model"* (Pirozzi, 2017). Basically, we can identify three main domains:

- *the generated value, which corresponds to the deliverables, and which tends toward being visible;*
- *the perceived value, which corresponds to stakeholder satisfaction, and which tends toward being hidden;*
- *the value of changes, which corresponds to project risks and changes to the project, and which tends toward being partially both visible and hidden.*

During project life cycle, these three domains of value, and of correspondent results, evolve incorporating added value, and continuously both interact with each other, and affect each other, and are affected by each other, and this value stream results, at the completion of the project, in the release of project delivered value (see Figure 9.2).

The deliverables are the foundations of building project success, their purpose is achieving project objectives and, then, they are necessary but not sufficient to the achievements of project goals, which require stakeholder satisfaction; they are visible, and directly related with generated value. The development of the deliverables is based on project requirements, which project managers' community tends to consider as they are "objective" (although, since they are the result of a mediation among diverse "subjective" stakeholder expectation, they are not, as already mentioned in previous chapters and as it will be better detailed in the following), and it is driven by written rules and/or specifications, including contracts, Statements of Work, technical specifications and/or requirements, constraints, standards, and organizational structures. In any case, the deliverables are objectively measured in terms of consistency and/or quality, cost, and timing, and, then, they are oriented mainly to efficiency; project information is mainly required.

The stakeholder satisfaction is based on stakeholder expectations, it requires the existence of proper deliverables, but its purpose is achieving those project goals that have been the justification of mutual investments, and, therefore, as we will further in following chapters, it turns out to be the critical success factor in all projects. Stakeholder satisfaction is mostly hidden, and/or tends to be hidden, and it is directly related with perceived value. Moreover, stakeholder satisfaction is a subjective issue that is generally subjectively measured, but it is absolutely oriented to efficacy; it is also behavior-driven, and, since it is based on relations, it requires interactive communications.

Finally, on the "waterline", there are those risks (including of course both the opportunities and the threats) and those changes, which are both based on events, and which may

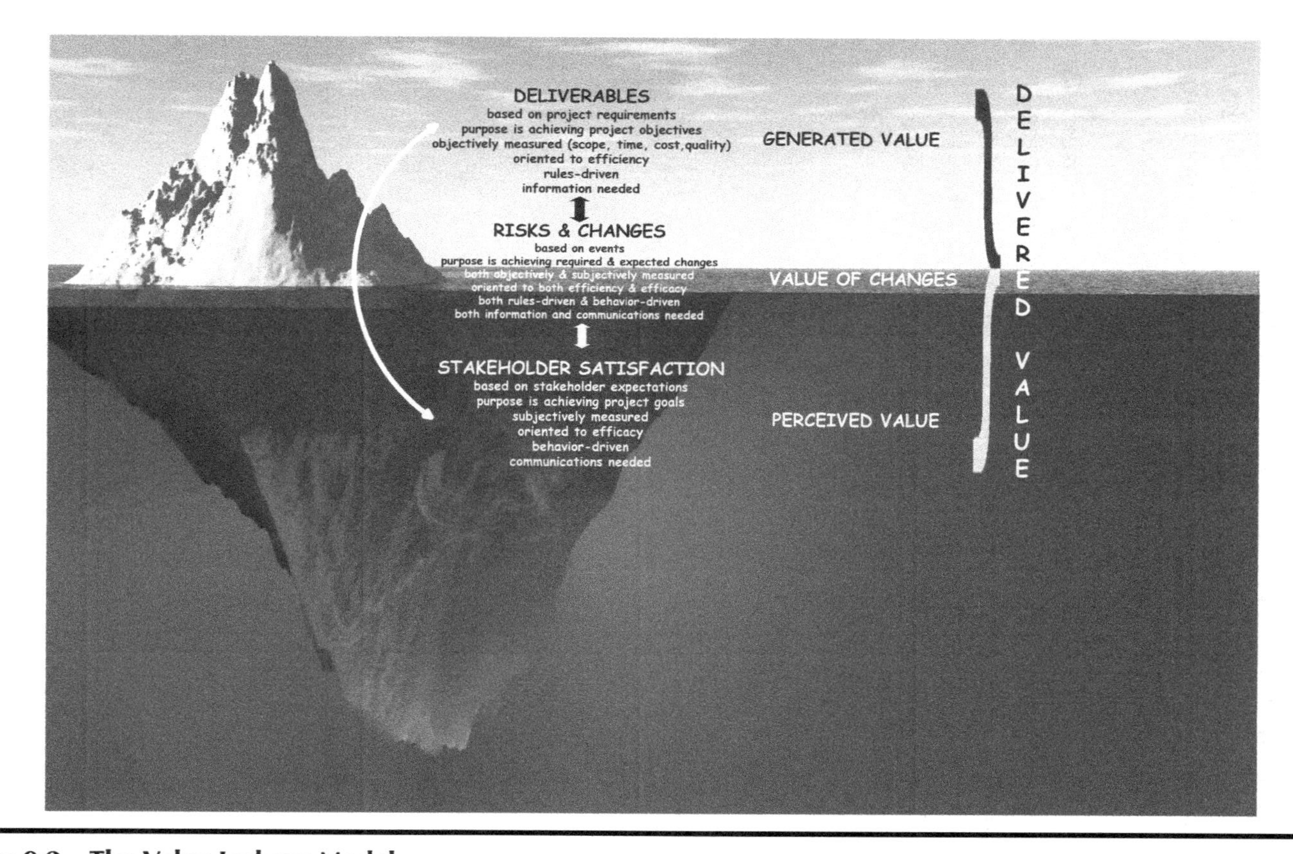

Figure 9.2 The Value Iceberg Model.

significantly influence and/or have impact on both deliverables and the stakeholder satisfaction. Generally, risks and changes have to be measured both objectively and subjectively, they are oriented to both efficacy and efficiency, they are both rules-driven and behavior-driven, and they require both project information exchange and the development of interactive communication.

Definitively, due to this indissoluble link between stakeholder relations and project delivered value, satisfying both stakeholder requirements and stakeholder expectations becomes critical for achieving project success.

Chapter 10

Satisfying Stakeholder Requirements and Expectations: The Critical Success Factor

If we consider present international and national standards for project management, may be one of the most updated definitions of project management is "the application of knowledge, skills, tools, and techniques to project activities to meet the project requirements", together with "*effective project management helps individuals, groups, and public and private organizations to meet business objectives (and to) satisfy stakeholder expectations*" (Project Management Institute, 2017). *In general, although undoubtedly the attention to stakeholders, and specifically to stakeholder expectations, increased significantly in last years, the dichotomy between requirements and expectations led, and still leads, to harmful misunderstandings, and, moreover, it is clearly, as we will see further in this chapter, a basic cause for projects lacks of success and/or failures.*

In fact, *possible misunderstandings are due to the fact that, while we consider natural and/or normal that stakeholders have one-sided behaviors in accordance with the diversity of their interests, we tend to consider the projects as if they are neutral*; in other words, *while we feel uncomfortable when dealing with stakeholder expectations, because they are subjective, we feel more comfortable when dealing with project requirements, which we tend to consider as objective ... although they are intrinsically not.* Actually, *project requirements are nothing but stakeholder requirements*, and, moreover, *requirements are the result of a complex, non-linear, and affected by semantic noise mediation among diverse subjective expectations*, which, although it has been initially somehow agreed when stakeholders signed the contract, can be evidently interpreted differently by different "stakeholders at stake". Furthermore, *requirements have a dynamic nature too*, and this can be either positive and/or neutral for the project, if they are managed accurately and properly during all project life cycle, or negative, if they diverge from what it has been "apparently" agreed before so that, in most of these cases "scope creeps" phenomena arise, or, definitively, as it happens quite often, a combination of positivity and negativity.

In fact, as previously mentioned, organizations define strategies, which are based on their own mission and vision, then select opportunities in accordance with defined strategy, then set business cases up, and, finally, start projects up. The inputs of a project, and, specifically, to the project management Initiating process group, include business case, contract, and Statement of Work (The International Organization for Standardization, 2012); generally, of course, there are different business cases for different stakeholders, as, for instance, providers and customers are. While the business cases, which are the originators of project start-ups, are based on stakeholder business expectations, whose satisfaction corresponds to the achievement of project goals, both the contract and the Statement of Work, which are the references for project development and delivery, are based on stakeholder requirements, which are, in turn, the conversion of different stakeholder expectations in a commonly agreed (at least initially) project scope, and fulfillment of which corresponds to the achievement of project objectives.

Does this conversion from stakeholder expectations to project/stakeholder requirements work effectively? Moreover, can the satisfaction of project requirements be considered an outcome that is sufficient to ensure the project success? The actual information that is available from the field (Project Management Institute, 2018) absolutely confirms the answer "no, not at all", although the benefits due to an increasing Project Management Maturity in the Organizations are evident. In fact, on average:

- *more than 30% of projects do not successfully meet those original goals and business intents on which their existence itself is based on, i.e., they do not satisfy stakeholder expectations*;
- *more than 50% of projects experience scope creep or uncontrolled changes to the project's scope, i.e., they do not satisfy original project requirements, which tend to grow abnormally during project life cycle*;
- *almost 50% of projects do not finish within their initially scheduled times, i.e., they do not satisfy original project time requirements*; and
- *more than 40% of projects do not finish within their initial budgets, i.e., they do not still satisfy original project cost requirements.*

Above-mentioned evidence of projects' not-so-brilliant performances—can we imagine what it would happen if going to a potential investor and asking him to fund a project while showing average performances like above?—lead us to an important principle: *stakeholder perspective is a definite driver for project success, and, even if it includes the subjectivity of relations, it is more reliable, and controllable, than traditional stand-alone project requirements perspective, which, in any case, is objective only apparently, since requirements are a mediation of different subjective stakeholder expectations.*

Indeed, *a project is really successful when its results, in terms of delivered value, do not only achieve those project objectives that traditionally correspond to the fulfillment of project requirements, but are also perceived as they will achieve those project goals, which correspond to the satisfaction of stakeholder expectations; perception becomes a basic driver during project life cycle,* because project's performances could be evidently measured only after project completion, i.e., during following product/service/infrastructure life cycle, and, then, subjectivity of stakeholder relations takes, through and through, that central role, which is crucial for driving stakeholder satisfaction. *In fact, depending on the complexity of the projects, the gap between the generated value in terms of deliverables, and the perceived value in terms of satisfaction, may become significant,* as we will see in next chapter.

In the stakeholder perspective, starting from the invested value in terms of resources, stakeholder relations both generate a value in terms of deliverables, and determine a perceived value in terms of satisfaction. Indeed, a *rational domain of requirements,* which is oriented to efficiency and targets project objectives, *interacts with a relational domain of expectations,* which is oriented to efficacy and targets project goals: *each of the two domains both influences and supports the other, and a stream of project value is created to be delivered at project completion* (see Figure 10.1).

Definitively, effective stakeholder management should target both the fulfillment of project/stakeholder requirements and the satisfaction of stakeholder expectations, which correspond to both the achievement of project objectives and the perception that project goals will be achieved: stakeholder satisfaction, instead of being "a" critical success factor, proves then to be

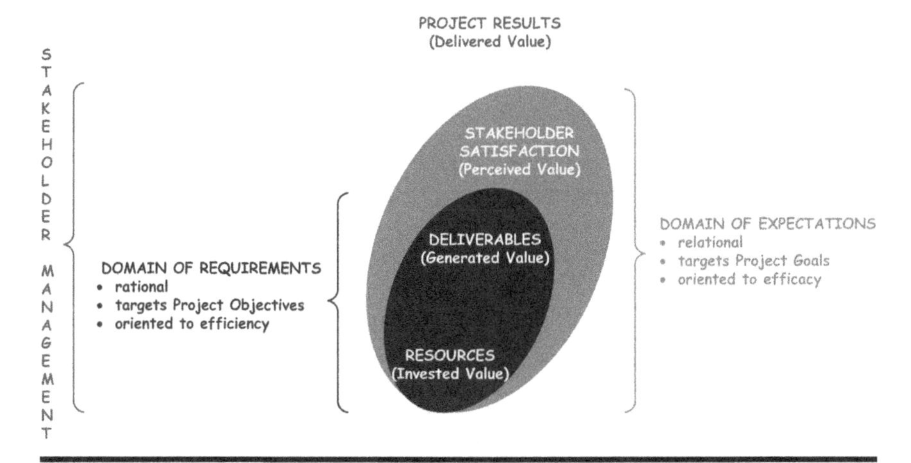

Figure 10.1 The stakeholder perspective.

"the" critical success factor; in fact, projects may not succeed their goals, or may fail at all, for various reasons, which could be technically very different, but, for sure, each project that was not successful had at least one key stakeholder whose expectations were not satisfied (Pirozzi, 2017).

Chapter 11

Facing Successfully Different Levels of Project Complexity

Each project is characterized by an inherent degree of multi-level complexity; this is mainly because, on the one side, all projects are unique and unrepeatable, while, on the other side, all their relations are intrinsically non-linear, and these attributes are valid not only for the projects themselves as a whole, but also for their results, deliverables, scope, time, cost, quality, value, constraints, assumptions, risks, and, of course, stakeholders. *Therefore, the discipline of project management arises to be not only useful to meet project requirements with increasing efficacy and efficiency, but also foundational to face project complexity*, although it seems that this advantageous aspect is not commonly, and somehow unexpectedly, highlighted.

Ultimately, *project management faces project complexity both rationally and relationally. The rational approach is the most well-known and mature part of project management*, and it is also its most common understanding; *rationally, the complexity of the project is addressed*:

- *by providing proper breakdowns that lead to the possibility of both handling and estimating more easily lower complexity work packages, which will then be integrated bottom-up to provide appropriate project outcomes and deliverables*;
- *by making efficient use of Initiating, Planning, Executing, Monitoring and Controlling, and Closing process groups, and this also by making an effective process tailoring, and, especially, by establishing an appropriate baseline as reference for project development, and, then, by properly controlling project's progress*;
- *by developing a specific, and transparent, team-based project organization, which can greatly increase both efficacy and efficiency, and, at the same time, can drastically reduce eventual endogenous complexity due to the permanent organizational structure's constraints and indeterminacies*; and
- *by managing risks, and by doing it properly*, in order to taking the uncertainty of the context and/or of the project environment into account.

Relational issues are of extraordinary importance in facing project complexity. First aspect to be considered is that *stakeholder relations are themselves characterized by a multilevel complexity*, since, as previously stated:

- stakeholders are persons, or groups of persons, i.e., complex systems;
- stakeholders are different, they may speak or understand different languages, and they have different interests;
- stakeholders are numerous, and stakeholder relations are even more numerous;
- stakeholder relations are context dependent, and they influence each other;

- all stakeholder relations are important, and, at least, have to be monitored; and
- stakeholder relations may be evolutive.

Therefore, since stakeholder relations introduce complexity, it should be evident, although sometimes it seems that is not, that this *relational complexity could and should be faced, and possibly solved, relationally only*, i.e., through the same relations that generate the problem: in other words, *it is inconceivable that a relational complexity can be solved rationally*, e.g., *through a better planning and control.* However, it is basic that *stakeholder relations have a primary supportive role*, and not only a role of complexity bringers, and, moreover, that *their positive importance and usefulness is even more than above and that it is still increasing.*

A second relational issue that is of extraordinary importance in facing complexity is *teamwork, since it is not only the major factor for creating value, but it is also the major factor of destruction and removal of endogenous complexity, so generating a huge regaining and/or increase of productivity in the project. The virtuous circle is due to the emerging of "warm" individual responsibilities with respect to "cold" organizational responsibilities, which in project management takes place in Planning process group, and, specifically, when assigning transparently, after both work breakdown structure and project organizational breakdown structure have been assessed, the responsibilities of different work packages to diverse team members. From this point on, indeed, project team stakeholder relations evolve to efficient professional relations, since they succeed to overcome the constraints due to the existence of permanent hierarchical and/or functional organizational structures, which, being based on functions to be held and/or maintained rather than on objectives to be achieved, generally introduce endogenous complexity with those uncertainties about "who is the owner of what", and/or "who is doing what" and/or "waiting something from somebody", which almost always generate huge loss of time and cost.*

Ultimately, *a major relational approach in facing complexity is stakeholder perspective at all: key processes as development and management of stakeholder relations, management of stakeholder requirements and expectations, and stakeholder engagement, can greatly support project success, and it may be useful, in order to identify properly the most efficacious actions, to assess the different levels of complexity that may characterize the diverse projects* (Pirozzi, 2018).

A model which can be very helpful to face complexity, by supporting effectively decision-making processes, is the well-known Cynefin Framework, which have been created, and developed, starting from early 2000s (Snowden, 2010). Cynefin Framework is properly a sense-making model based on observation, in which "data precede model", rather than a theoretical categorization model, in which "model precedes data": *it individuates four domains which are characterized by different levels of complexity, i.e., simple, complicated, complex, and chaotic.* The domains of simple and complicated are regarded as *ordered.* In the simple domain, cause and effort relationships exist, are predictable, repeatable, self-evident, and can be determined in advance. The decision model is based on the unique best practice, and the most appropriate action path is sense–categorize–respond; communication is not present, only information is broadcast, and the network of relations is irrelevant. On the other hand, in the complicated domain, cause and effort relationships exist, but are not evident; the right answer requires the use of analytical methods, and the support of experts. The decision model is based either on one of the good practices or on a combination of them, and the most appropriate action path is sense–analyze–respond; communication is mainly informative, and the network of relations starts to be important. The domains of complex and chaotic are regarded as *unordered.*

In the complex domain, cause and effort relationships are only obvious in hindsight, i.e., retrospectively, with unpredictable, emergent outcomes. The decision model is based on an emergent practice, and the most appropriate action path is probe–sense–respond; communication is mainly interactive, and the network of relations becomes fundamental. On the other hand, in the chaotic domain, no cause and effort relationships can be determined, and quick actions that are finalized to target more stability are necessary. The decision model is based on a novel practice, and the most appropriate action path is act–sense–respond; also in this domain, communication is mainly interactive, and the network of relations is fundamental. Finally, it is very interesting to consider the peculiar simple/chaotic boundary, since *a "shortcut towards the chaos" may occur in possible dynamics: i.e., if we push to oversimplify systems, which, in any case, are either complicated or complex, and we try to manage them as if they are "simple", domain will change very soon into an unmanageable chaos.*

Since the operational processes, including projects and operations, may be seen as the causes, and the generated value as the effect, Cynefin Framework *can be also applied to projects* (see Figure 11.1), and, then, it can help to manage projects in contexts of different levels of complexity, in which we look for the causes that could lead to the effects of projects successes, as well as for the most effective project management actions to be taken (Pirozzi, 2018).

In the ordered domains, actually, as we could expect, *no project belongs to the Simple domain, which is the domain of operations.* In this domain, the value generation is a consequence of achieving the target of operation objectives: normal operations management is sufficient, but procedures are essential, and *following procedures is the critical success factor*; there is just one lesson-to-be-learned, which is the unique best practice to follow. In the complicated domain, we find, here also as we could expect, *complicated projects*; support of experts

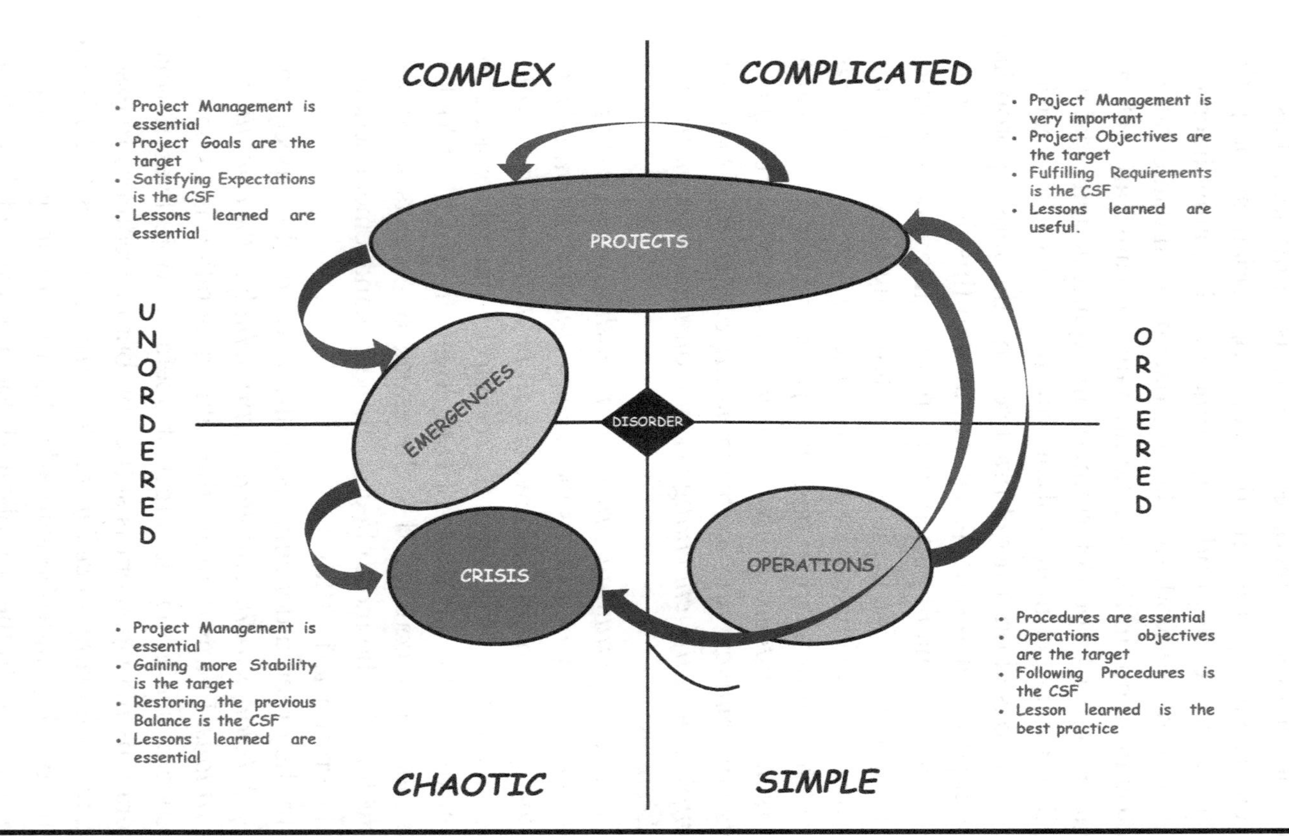

Figure 11.1 Projects in Cynefin Framework.

is needed, and that is where project managers start to come in. In this domain, the value generation is a consequence of achieving the target of project objectives; the discipline of project management is very important and/or essential, and *fulfilling the project requirements is the critical success factor*; lessons learned about good practices are very useful.

In the unordered domains, we find in complex domain, as we could expect also in this case, the *complex projects*, but *also a part of "manageable" emergencies*, which include those events that are substantially unpredictable, but where it is possible to plan adequate responses, and/or to manage properly the relevant risks. In this domain, the value generation is a consequence of achieving the target of project goals; the discipline of project management is essential, and *satisfying the stakeholder expectations is the critical success factor*; lessons learned are essential to find out a proper emergent practice. Finally, in the chaotic domain, there are both the "unmanaged" emergencies and the *crisis*. In this domain, the value generation is a consequence of achieving the target of gaining more stability; the discipline of project management is essential, and *restoring the previous balance is the critical success factor*; lessons learned are essential also in this domain to find out a proper novel practice.

It may be interesting to notice that *eventual poor management of complexity leads to a "counterclockwise" effect that makes complexity itself worse*; poorly managed operations become complicated projects, poorly managed complicated projects become complex projects, poorly managed complex projects become emergencies, and poorly managed emergencies become crisis. *In addition, if a project is oversimplified in an operation, the Cynefin's "shortcut toward the chaos" will generate very soon a "snowball effect" in the direction of an unmanageable chaos.*

Going back to *complicated and project projects, the stakeholder perspective can help to understand in depth their differences* (Pirozzi, 2018), and also those historical dichotomies as traditional versus agile, value-driven versus plan-driven, project management 2.0 versus 1.0, etc. (see Figure 11.2).

In complicated projects, there is a small gap between meeting the requirements and meeting the expectations of stakeholders, and this happens when:

- *the project is part of the customer's core business* ("supplier perspective", as in internal or in outsourcing projects), and, then, for the customer, the project is the business goal, and/or project results are product-oriented and/or tangible (e.g., in some infrastructure projects) and/or, in any case, stakeholder requirements are either well-defined (as in traditional contexts) or are evolutionary, but, in both cases, all stakeholders cooperate effectively (as in agile contexts);
- *projects are essentially plan-driven*;
- *the triple constraints Time-Cost-Quality are dominant*; and
- *the relations with stakeholders are important and periodic.*

Since, in complicated projects, the domains of the stakeholder expectations and of the stakeholder requirements substantially overlap, we can assume that success is based on the fulfillment of stakeholder requirements, and that, therefore, managing properly the generated value, whose measures consist, as in classic project management, in cost, time, earned value, and, generally, in consistency/progression of the deliverables, is necessary and sufficient.

On the other hand, *in complex projects, there is a significant gap between meeting the stakeholder requirements and meeting the stakeholder expectations* (see Figure 11.2), and this happens when:

- *the project is a support of the customer's core business* ("customer perspective", as in most external projects), and, then, for the customer, the project is a mean to achieve

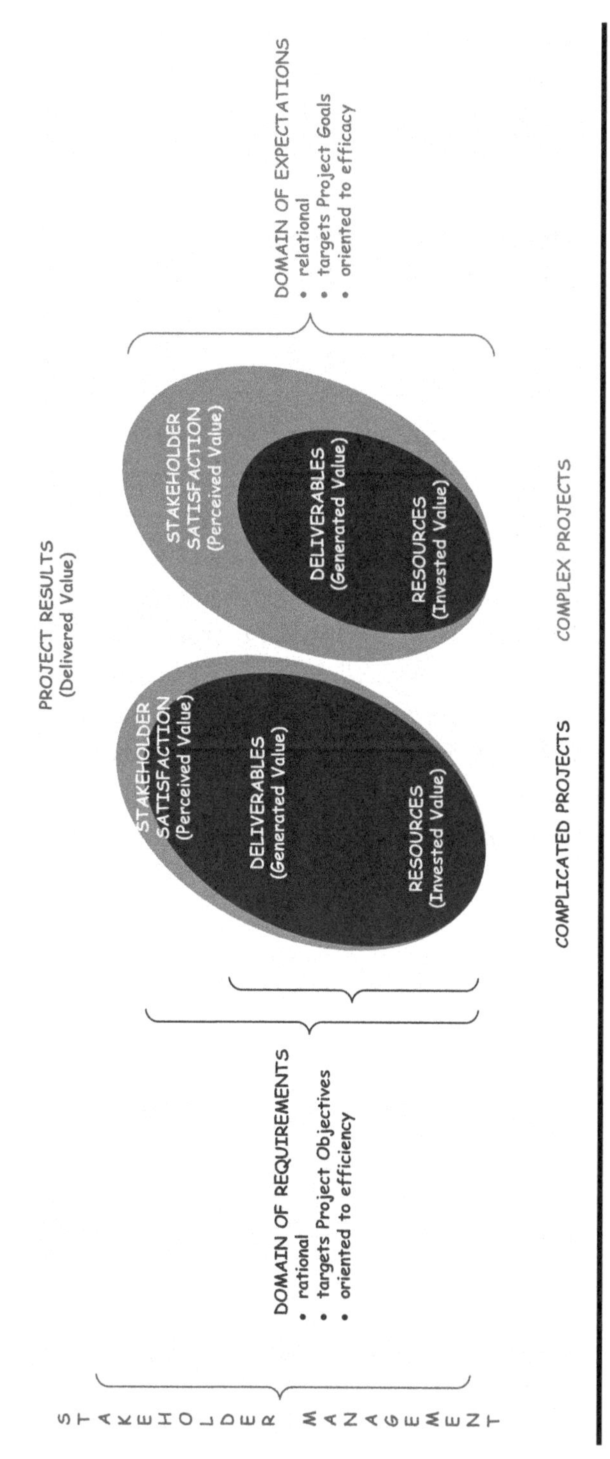

Figure 11.2 The stakeholder perspective in complicated and complex projects.

his business goals, and/or the project results are service-oriented and/or intangible (e.g., in software projects), and/or stakeholder requirements are not well-defined, or are evolutionary, but not all stakeholders cooperate effectively;

- *projects are essentially value-driven*;
- *competing constraints are dominant: the importance of value and reputation is superior to that of the triple constraints* (Kerzner, 2015); and
- *relationships with stakeholders are primary and can be continuous, fast, interactive (as in 2.0 world), evolutive* (Kerzner, 2015).

Since, *in complex projects,* expected project goals can be far away from required project objectives, *the project success is based on the satisfaction of stakeholder expectations, and, therefore, managing properly the perceived business value becomes mandatory. However, successful management of the business value requires adequate metrics and measures, which can be used also during project life cycle, and not only after project completion; proper key performance indicators are then needed.*

Chapter 12

Targeting Both Project and Business Value Generation by Using KPIs

In all projects, which are, in any case, either complicated or complex, success derives from the capability of satisfying stakeholders by both generating the required project value and delivering the expected perceived business value. "Success is not necessarily achieved by completing the project within time, cost, and scope. Success is when the planned business value is achieved within the imposed constraints and assumptions" (Kerzner and Saladis, 2009). However, managing value requires adequate metrics and measures (Kerzner, 2017); moreover, since the expectations of different stakeholder communities are evidently diverse, the relevant measures and estimates must include a set of parameters that cover project management, economic, and business and/or social needs (Pirozzi, 2018), and, then, cannot be limited to earned value, which is powerful and extraordinary, but is unavoidably based on requirements only, and not on expectations too.

Project/stakeholder requirements, which reflect diverse stakeholder expectations but do not generally coincide with any of them, are the basis, during project life cycle, of those measures of actual cost, and evaluations of the state of progress, which are traditionally used as indicators to estimate the "present" situation and the "future" time and cost of the project completion. In any case, but especially in the case of complex projects, metrics and indicators that are based on earned value could effectively be integrated by other useful project management indicators that can be objectively measured (Kerzner, 2017). *On the other hand, generated business and/or social value, which could represent effectively a measure of the satisfaction of stakeholder expectations, is, in any case, future with respect to project life cycle, and, unfortunately, can be measured only after project completion; however, appropriate indicators that are generally used to objectively measure performance during product/service life cycle can be very effectively innovatively used during project life cycle too as subjective measures of perceived business and/or social value (see Figure 12.1), and, then, provide a powerful support to either confirm or readdress, in terms of both deliverables and stakeholder satisfaction, the action of the project team during the entire project life cycle (Pirozzi, 2018).*

Definitively, during project life cycle, we therefore need the support of other additional indicators that can represent effectively both the project value and the business/social value. Proper *KPIs* (Parmenter, 2015) *are, therefore, very useful and/or necessary to target the success of both complicated and complex projects* (Pirozzi, 2018). In fact, when managing major projects, KPIs are, in any case, part of the necessary multidimensional evaluation of project success and value (Archibald and Archibald, 2016). The same consideration can be applied in case of managing complex projects, too. *Indeed, although KPIs are fundamental measures of released projects/products/services, KPI-based measures and estimations can be also extremely useful to get crucial progress indications about the*

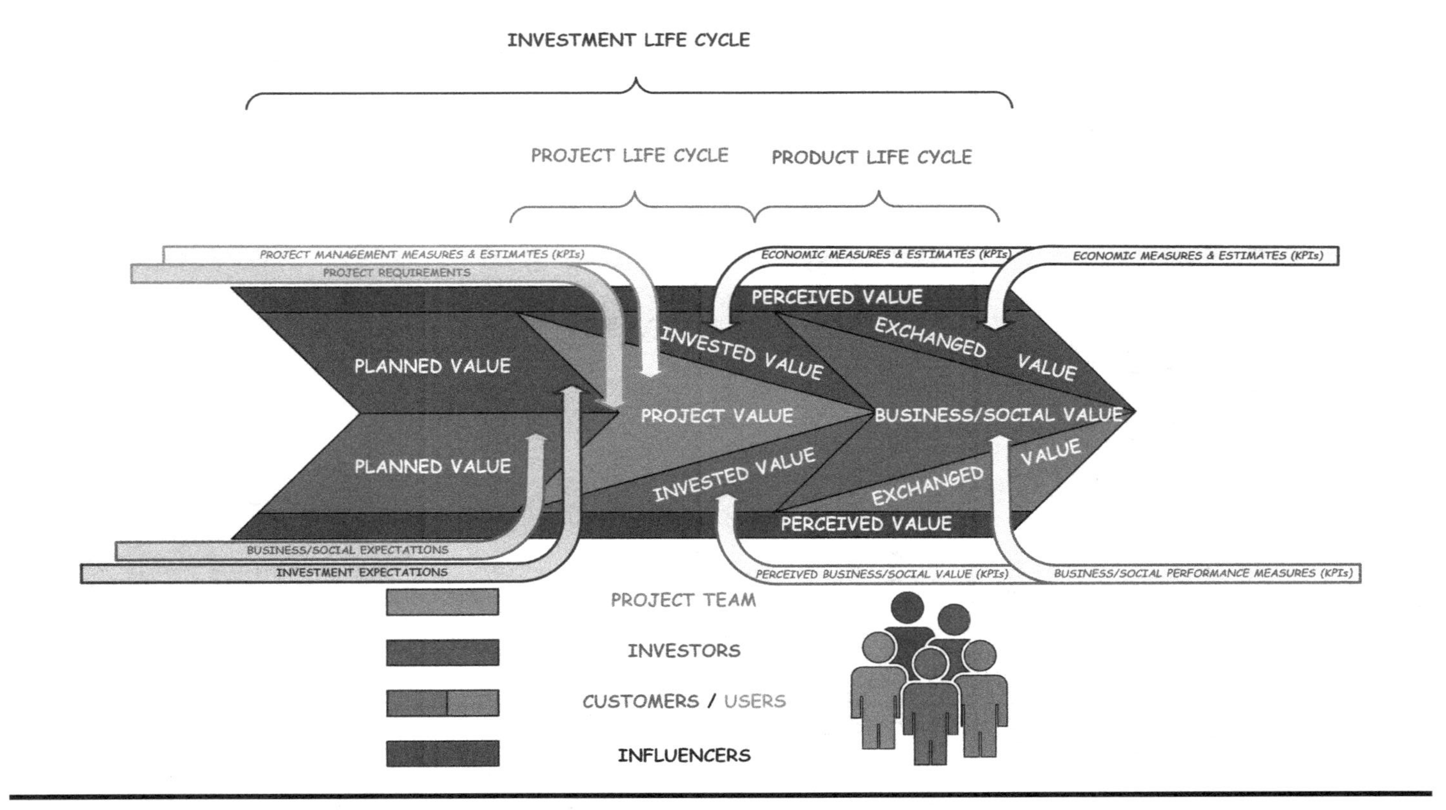

Figure 12.1 A value chain of whole investment life cycle and relevant project KPIs.

generated value, and to monitor stakeholder expectations, during investment, project, and product/service life cycles.

However, project stakeholders are different, have a different behavior, which characterize them in communities, and, then, have different expectations (Pirozzi, 2017); stakeholder communities, indeed, target different types of value. *Providers* (project manager, project team, business partners) *target a technical (delivered) value*, which include triple constraints, project objectives, revenues, while *Investors* (top management, shareholders, funders…) *target an economic value*, which include costs, revenues, business prospects, and *Purchasers* (customers, users) *target their business value*, including customer costs (which, of course, coincide with providers/investors revenues), project goals, expected benefits achievement. Specifically, *KPIs should address different types of value to cover both project management, economic, and business domains.*

Project Management KPIs are especially useful to enhance project control and to maintain and/or modify the proper route toward deliveries that fulfill stakeholder requirements. They are very helpful both in complicated and in complex projects. Main example of these KPIs is Earned Value, which, although it is almost never defined as a KPI, is used in almost all projects, and it is often the only KPI which is used in traditional, classic, or complicated Project Management. Additional KPIs include, for example (Kerzner, 2017):

- cost variance;
- schedule variance;
- cost performance index;
- schedule performance index;
- resource utilization;
- percent of milestones missed;
- management support hours as a percent of labor;
- planning cost as a percentage of labor;
- percent of assumptions that have changed;

- percent turnover of key workers;
- percent of work packages adhering to the schedule and/or to the budget;
- number of assigned resources versus planned resources;
- percent of actual versus planned baselines completed to date;
- percent of actual versus planned best practices used;
- number of critical assumptions made;
- percent of critical assumptions that have changed;
- number of cost and/or schedule revisions;
- number of critical constraints; and,
- percent of work packages with a critical risk designation.

Economic KPIs are especially useful to improve relations with top management and funders, and to maintain and/or modify the proper route toward the satisfaction of their economic and financial expectations: their use can be very helpful in complicated projects, and it is basic in complex projects. In any case, since the domain of economic KPIs is very analytical, and very vast, it is preferable to narrow focus on some selected high-level indexes. Economic KPIs include, for example (Marr, 2012):

- Economic and Financial Indicators, as Net Profit, Net Profit Margin, Gross Profit Margin, Operating Profit Margin, EBITDA, Revenue Growth Rate, Total Shareholder Return, Economic Value Added, Return on Investment, Return on Capital Employed, Return on Assets, Return on Equity, Debt-to-Equity Ratio, Cash Conversion Cycle, Working Capital Ratio, Operating Expense Ratio, CAPEX to Sales Ratio;
- Marketing Indicators, as Market Growth Rate, Market Share, Brand Equity, Cost per Lead, Conversion Rate, Search Engine Rankings and click-through rate, Page Views and Bounce Rate;
- Customer Relationship Management Indicators, as Net Promoter Score, Customer Retention Rate, Customer

Satisfaction Index, Customer Profitability Score, Customer Lifetime Value, Customer Turnover Rate, Customer Engagement, Customer Complaints;
- Human Resource Indicators, as Human Capital Value Added, Revenue Per Employee, Employee Satisfaction Index, Employee Engagement Level, Staff Advocacy Score, Average Employee Tenure, Absenteeism Bradford Factor, 360-Degree Feedback Score, Salary Competitiveness Ratio, Time to Hire, Training Return on Investment; and
- Sustainability Indicators, as Carbon Footprint, Water Footprint, Energy Consumption, Saving Levels Due to Conservation and Improvement Efforts, Supply Chain Miles, Waste Reduction Rate, Product Recycling Rate.

Ultimately, Business Value KPIs are either common or specific for each sector of activity. They are especially useful to improve relations with customers and users, and to maintain and/or modify the proper route toward the satisfaction of their business expectations; their use is foundational in complex projects (Pirozzi, 2018). The business value KPIs that are common to the different sectors of activity are of primary importance, since they include

- *measures and percentages of stakeholder satisfaction (in terms of both requirements and expectations);*
- *measures and percentages of stakeholder positive engagement;* and
- *measures of perceived value, as perceived business value, social value, quality, reputation, business climate, innovation, sustainability.*

In addition, there are other business value KPIs, as functional and/or quantitative measures, and the relevant percentages of completion/deviation from budget/schedule, which are specific of each sector of activity; some examples of these KPIs and/or KPI groups are shown below:

- *Sustainable Smart Cities* (International Telecommunication Union, 2015): indices of ICT, environmental sustainability, productivity, quality of life, equity, social inclusion, physical infrastructure, network, knowledge economy, education, openness and public participation, governance, infrastructure connections;
- *Railway Infrastructure* (PRIME, 2017): measures of context, safety/security/environment, performance (punctuality & robustness), delivery (capacity & condition), finance (costs, charging, & revenues), growth (utilization, intermodality, & asset capability/ERTMS);
- *Construction Industry* (Glenigan, Constructing Excellence, CITB, and UK Department for Business, Energy, and Industrial Strategy, 2018): Economic KPIs (Client Satisfaction, Contractor Satisfaction, Defects Predictability Cost, Predictability Time, Profitability, Productivity), Respect for People KPIs (Staff Turnover, Sickness Absence, Safety, Working Hours, Qualifications & Skills, Training, Investors in People, Staff Loss, Make-up of Staff), Environment KPIs (Energy Use, Mains Water Use, Waste, Commercial vehicle movements);
- *Local Public Transportation* (Trans Cooperative Research Program, 2003): measures of availability, service delivery, community, travel time, safety and security, maintenance and construction, cost, capacity;
- *Pharmaceutical Industry* (Roche, 2017): measures relevant to value for patients, for employees, for partners, for communities, for environment, for investors;
- *Research Infrastructures* (ESFRI, 2018): indicators relevant to scientific excellence outcomes, output and delivery of talent, reference role in the disciplinary field (uniqueness in capabilities, in capacity) at International level,

progress in achieving the Research Infrastructures' milestones along its life cycle, impact, innovation, entrepreneurship, establishment and development of the Research Infrastructures' user community, scientific data management policy, metadata catalogue interfacing, open science initiatives, advanced data services for scientific analysis and for innovation developments, enforcement of quality control of access (peer review), data, and services to research and innovation;

- *Web Marketing* (these KPIs are derived from Google Analytics Software): measures of audience (number of sessions, users—both new and returning, page views, average session duration, bounce rate, new sessions), location, new versus returning users, browsers and OS, devices, acquisition (direct traffic, organic search, referral, social media, display advertising, email, paid search), source/medium (search engine/domain, organic/cost-per-click paid search/referral), Ad Works/SEO/Social, behavior (pages, actions, number of page views, bounce rate, exit rate, flow), site content/all pages, landing/exit pages, real-time, in-page analytics;
- *Information Technology* (Kerzner, 2017): number of lines of code, language understandability, information movability/immovability, software complexity, math/algorithms complexity, input/output understandability, percent of the system with latest updates (antivirus and antispyware), mean time to make system repairs; and
- *Roadway Bridges* (European Commission—COST Action TU1406, 2016): KPIs at Component Level (Substructure, Superstructure, Road plus Equipment), at System Level (Technical, Sustainability, Socio Economic), at Network Level, and correspondence with Maintenance Tasks.

In project management, any measurable value could be effectively used as a KPI, but, if they are too many, the situation can become unworkable: it is, then, important to make an appropriate selection. While *all KPIs have to be in accordance with SMART rule* (Doran, 1981), i.e., specific, measurable, attainable, realistic or relevant, time related, *project-oriented KPIs have to be selected also because they have the characteristics of being predictive, measurable, actionable, relevant, automated, and few in number*: moreover, *the best way to share effectively, rapidly, and continuously, KPIs with other stakeholders is using dashboards and/or scorecards, which replace very efficiently traditional reports* (Kerzner, 2015). As previously seen, the use of dashboards, in addition, can bring important benefits in the relations with two specific typologies of non-positive stakeholders, as reluctant/indifferent stakeholders, who do not want to be engaged, and negative/hostile stakeholders, who do not want to agree anything: these categories are, unfortunately, quite common in the real world, and especially in complex projects. Actually, the use of KPIs and dashboards can help to deal effectively also with these stakeholders, because *KPIs that are shared via dashboards are business-oriented, client-centered, and very stakeholder-friendly; moreover, they require a quick and minimal effort to interact, and, in most cases, they are available so frequently for sharing, that also no-answers can be interpreted positively, as a silent approval.*

In all cases, KPIs are necessary, powerful, and effective means to manage both delivered and perceived value: proper KPIs can be, therefore, selected, agreed, measured/estimated, shared with stakeholders via dashboards, and used to confirm/readdress, in terms of both deliverables and stakeholder satisfaction, the action of the project team during the entire project life cycle (Pirozzi, 2018). *In this way, an effective stakeholder management, which uses properly KPIs and dashboards, can increase the success rate in a variety of projects with different size and complexity, by supporting both the value generation, and the project goals achievement.*

Chapter 13

Relationship Management Project: A Structured Path to Effectiveness

Whatever the approach to project management is, Stakeholder Relationship Management is absolutely cross-discipline; it concerns all relation management issues, and "relationship management is of special importance in today's world" (Archibald, 2017), but it has a fundamental indirect influence on delivery management too, since each project is made by stakeholders to be delivered to other stakeholders. In short, specifically (Pirozzi, 2018):

- *in complicated projects, Stakeholder Relationship Management supports the generation of that delivered value, which leads to project success by achieving project objectives, which, in turn, fulfill project/stakeholder requirements; therefore, Stakeholder Relationship Management has to be focused on those stakeholders who are involved in the implementation of the project, such as the project team, and its basic guidelines can be the*

development of awareness, leadership, teaming, motivation, and ethics (Pirozzi, 2017), to increase deliverable effectiveness and project efficiency; and
- *in complex projects, in addition to the above, Stakeholder Relationship Management supports the generation of that perceived business value, which leads to project success by achieving those project goals, which, in turn, fulfill stakeholder expectations; Stakeholder Relationship Management must, then, be primarily focused also on those stakeholders who direct the project, such as customers/users, sponsor/top management, and funders, and its basic guideline is the development of an effective communication, to add value to the project and to improve project effectiveness.*

Definitively, in all projects, but specifically in complex projects, the Stakeholder Relationship Management is a powerful, and effective, set of processes and competencies, that helps both the project manager and the project team to remain constantly aligned with both stakeholder requirements and expectations, in order to target continuously the achievement of both project objectives and goals, so increasing the overall project success rate (Pirozzi, 2018). However, in order to improve successfully both the meeting of the original goals and business intent of the project, and/or the achievement of objectives in terms of scope, time, cost, and quality, especially *in large and/or complex projects, just an event-driven stakeholder management cannot be considered as sufficient, and, therefore, a structured path to effectiveness must be built; every relationship management is, indeed, a project in the project, and it has to be managed properly.*

Relationship Management Project includes and integrates several specific enhanced project management processes, which, in turn, interact with each other perfectly in accordance with the five project management process groups, i.e., Initiating, Planning, Executing, Monitoring and Controlling, and Closing (see Figure 13.1); *in this way, immediate applicability is*

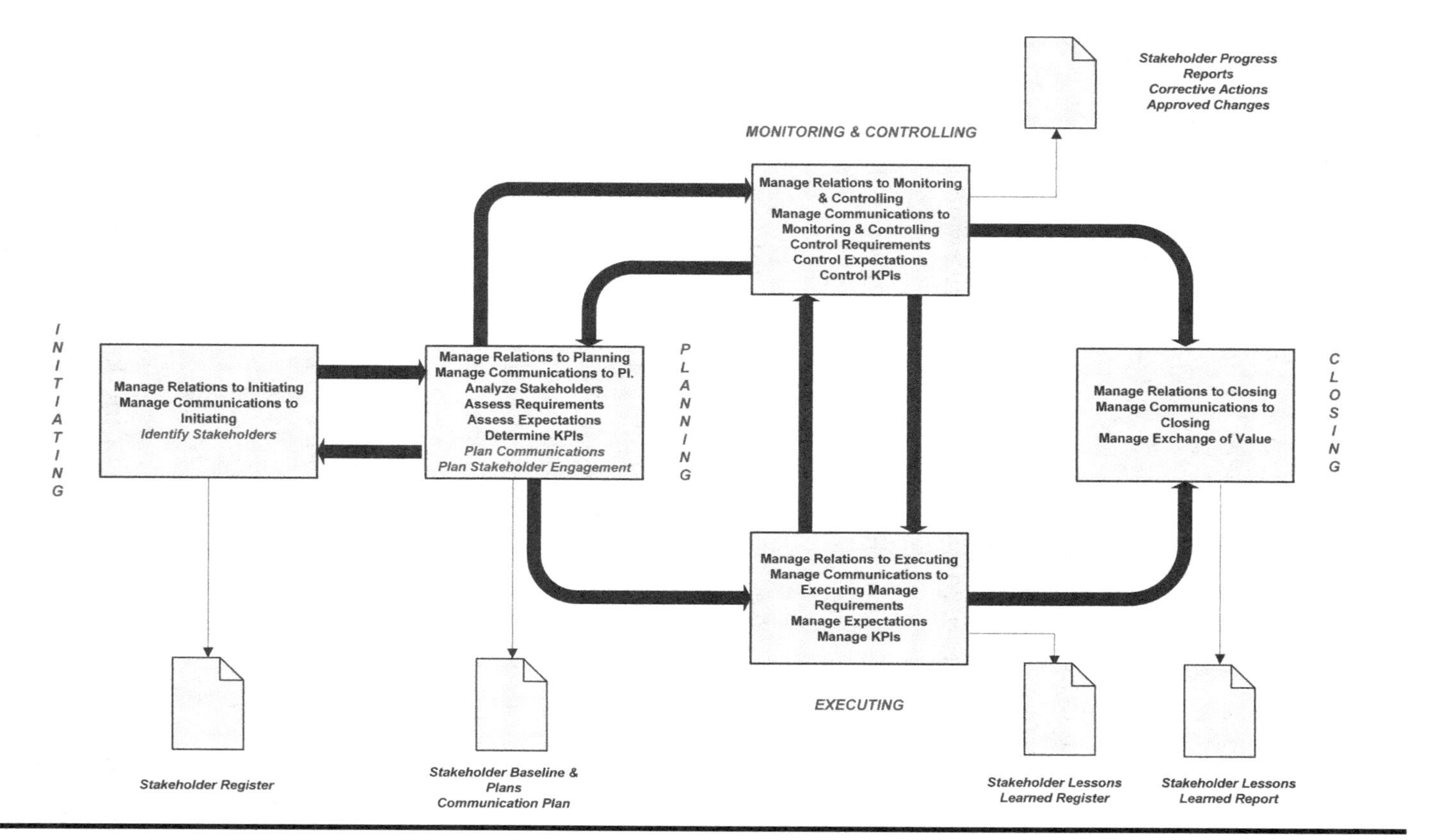

Figure 13.1 The Relationship Management Project.

guaranteed and ensured in all complicated and complex projects, of any size and in any sector of activity.

Relationship Management Project is naturally based on the axioms that both stakeholder relations management and communications have to be considered in their most comprehensive and extensive meaning, as we saw in previous chapters; indeed, *stakeholder management is not limited to stakeholder engagement only*, because we have to consider that, for instance, relations that are non-collaborative and/or among stakeholders, and not just with them, have to be managed too, as well as *project communication is not limited to project information only, but it must cover all the diverse multilateral exchange of project-related contents.*

Definitively, Relationship Management Project both is based on and follows the two main guidelines:

- *both stakeholder relations management and communication management processes must be specifically present in all the project management process groups*; e.g., it is evidently unconceivable to identify a scope or to develop a project charter without interacting with the project sponsor, or to develop project plans without interacting with those members of the project team that are the responsible of the diverse work packages, or to close a project without interacting with the customer, etc.; and
- *both the project/stakeholder requirements, the stakeholder expectations, and the measures of value as the key performance indicators (KPIs), have to be processed during the whole project life cycle, i.e., they have to be determined, assessed, managed, and control, and all this iteratively and/or adaptively too, if needed.*

In the Initiating process group of the Relationship Management Project, there are the following processes: *manage relations to initiating, manage communication to initiating, and identify stakeholders.* In this group, *peculiarity of both relations and communication management processes is precisely the complexity of initiating relations*, since there is the *forming* phase of diverse project stakeholder communities, and *there is also the delicacy due to importance of first impression, which can influence heavily, both positively and negatively, the continuation of the diverse stakeholder relationships*. The process *identify stakeholders* is almost identical to the namesake process that is present both in ISO 21500:2012 (International Organization for Standardization, 2012) and in *PMBOK Guide*, sixth edition (Project Management Institute, 2017). *Main document in output is the stakeholder register*, which is generally incorporated in the Project Charter.

In the Planning process group of the Relationship Management Project, there are the following processes: *manage relations to planning, manage communications to planning, analyze stakeholders, assess requirements, assess expectations, determine KPIs, plan communications,* and *plan stakeholder engagement*. In this group, *peculiarity of both relations and communication management processes is the complexity of navigating in newborn relations with diverse stakeholder community to reach significant "strategic" goals in terms of both scope, requirements, and expectations assessment, and planning accuracy and/or effectiveness. Analyze stakeholders process enhances traditional analysis issues through systemic approach, and, then, becomes foundational to establish the proper "stakeholder baseline" in terms of requirements, expectations, and KPIs. The processes of both assessing project/stakeholder requirements, which include the process collect requirements of* PMBOK Guide, *sixth edition (Project Management Institute, 2017), of assessing stakeholder expectations, and of determining KPIs, are all the results of an in-depth analysis, and in some cases of a detailed design too, and integrate hypotheses made by project team with interactive validations to be obtained through feedbacks originated by diverse stakeholders communities. Plan communication* is almost identical to the namesake process that is present both in ISO 21500:2012 (International

Organization for Standardization, 2012) and in *PMBOK Guide*, sixth edition (Project Management Institute, 2017), while *plan stakeholder engagement* is almost identical to the namesake process that is present in *PMBOK Guide*, sixth edition (Project Management Institute, 2017). *Main documents in output are the stakeholder baseline, which include all details in terms of project/stakeholder requirements, stakeholder expectations, and value measures as KPIs, the stakeholder engagement plan, and the communication plan; all these documents may be incorporated in the project plan.*

In the Executing process group of the Relationship Management Project, there are the following processes: *manage relations to executing, manage communications to executing, manage requirements, manage expectations, and manage KPIs*. In this group, *peculiarity of both relations and communication management processes is the needed pragmatism of finalizing relations with all diverse stakeholder communities to the concrete effectiveness and efficiency of project deliverables.* Furthermore, *management of requirements, expectations, and KPIs is also finalized to realize, and to measure, those deliverables that are in accordance with various needs. Manage communications process includes the namesake process* that is present both in ISO 21500:2012 (International Organization for Standardization, 2012) and in *PMBOK Guide*, sixth edition (Project Management Institute, 2017), and *manage relations process includes manage stakeholder engagement process* of *PMBOK Guide*, sixth edition (Project Management Institute, 2017), *but both processes must be considered as covering the whole axioms that have been described above. Main document in output is the lesson learned register that is relevant to stakeholders and to the relations with them.*

In the Monitoring and Controlling process group of the Relationship Management Project, there are the following processes: *Manage Relations to Monitoring and Controlling, Manage Communications to Monitoring and Controlling, Control Requirements, Control Expectations, and Control KPIs.*

In this group, *peculiarity of both relations and communication management processes is the delicacy of relations with team members, since both effectiveness, efficiency, and accuracy of the control processes need that control itself is always managed, and perceived, as the result of a joint collaborative action, and never as a third-party inspecting audit.* The processes of *controlling both the requirements, the expectations, and the KPIs, target the deviations from the "stakeholder baseline", and relevant eventual changes to be approved. Manage communication process includes monitor/control communication process* that is present both in ISO 21500:2012 (International Organization for Standardization, 2012) and in *PMBOK Guide*, sixth edition (Project Management Institute, 2017), and *manage relations process includes monitor stakeholder engagement process* of *PMBOK Guide*, sixth edition (Project Management Institute, 2017), *but both processes must be considered, in this process group too, as covering the whole axioms that have been previously described. Main documents in output are the progress reports, the corrective actions, and the approved changes, which are relevant to stakeholders and to the relations with them.*

In the Closing process group of the Relationship Management Project, there are the following processes: *manage relations to closing, manage communications to closing, and manage exchange of value.* In this group, *peculiarity of both relations and communication management processes is the specific focus on both acceptance and transfer of delivered value, which make in this process group the importance of the relationship so great, that closing may be considered as the real "ultimate test" of the relationship itself, and, moreover, of the project as a whole.* Indeed, although it seems that in several projects the closing is considered almost a sort of "automatic" or "semiautomatic" process group (may be some/several project failures depend on that, too?), *the processes of managing the exchange of value are extremely tricky, from both the rational and the relational perspectives; in fact, before the "final" project decisions are made in terms of acceptance, transfer of property and/or responsibility, authorizations for payments, winding up of groups, change of work and/or of duties, etc., all types of fears and doubts tend to recur in stakeholder communities, and this must be managed very carefully and very delicately. Main document in output is the report of the lesson learned that is relevant to stakeholders and to the relations with them,* which may successively be incorporated in a *lessons learned repository* in order to be available to other projects.

Definitively, the Relationship Management Project is a structured path to effectiveness in increasing the success rate in all projects, but specifically in complex and/or large projects, and its usefulness and immediate applicability are proven and strengthened by its immediate complementarity, and integrability, with project management processes, too.

Chapter 14

New Stakeholder-Centered Trends: Project Management X.0

"We all live in a world of Project Management 2.0" (Kerzner, 2015). Indeed, interactive Web 2.0 makes powerful tools available to the community, and, in project management domain, there is also the possibility for differently located, and even virtual, project teams, of cooperating effectively via distributed collaboration; however, Project Management 2.0 is much, much more than this.

In fact, *Project Management 2.0* (Kerzner, 2015) *addresses a 2.0 domain of complex projects:*

- in which *the stakeholders may be very numerous, and/or distributed, and/or with diversified interests*;
- *success of which depends on generated business value*, and not only on time, cost, and objectives achievements, since "success is not necessarily achieved by completing the project within time, cost, and scope. Success is when the planned business value is achieved within the imposed constraints and assumptions" (Kerzner and Saladis, 2009);

- where *competing constraints as the value, the reputation and the quality, not only must be added to the constraints as time, cost, and risks, but become dominant, too*;
- *requirements* of which, *generally, are not well defined, but are flexible and evolutionary*; and
- in which *stakeholder relations are fast, continuous, interactive, and evolutive.*

Ultimately, Project Management 2.0 is an evolution of Project Management 1.0 (Kerzner, 2015):

- which is *business value-driven*, since the orientation toward customers requires measures that are focused on value, and not only on time and cost; relevant *key performance indicators*, which can represent appropriately the future, to be represented and continuously shared with project stakeholders via dashboards and/or scorecards, can be then effectively used;
- which is *flexible*, so that also agile approaches may be useful, and, moreover, it is characterized by participative and collaborative leadership, high-level project management competencies present in all project team, distributed planning and control, *in order to ensure the client-centered-flexibility that is needed to target project success*; and
- which has at its disposal *powerful web-based project management tools, to be used to enable the distributed collaboration, also with respect to virtual and/or not co-located teams.*

It is very important to notice that, since *Project Management 2.0* is focused on business value, it *can be applied not only to operational projects, but to strategic projects too*, and this characteristic someway has opened the way to a certain proposal of Project Management 3.0. In fact, a definition of Project Management 3.0 is "*Project Management 3.0 is business-driven project management using value creation with a heavy focus on building a portfolio of projects*" (Kerzner, 2019). Some Project Management 3.0 key issues concern:

- *new increased roles for both project managers and executives*;
- the *entry of the project management in the board room, also to support strategic planning*;
- the *need of intangible measures too*, including image/reputation, collaboration, commitment, customer satisfaction, stress level, teamwork, employee morale, and sustainability;
- the grow in use of *dashboards (especially of those that are internally developed)*; and
- the acknowledgment of the importance of both other values other than economical, e.g., cultural, behavioral, social, etc., *and of soft skills.*

Both Stakeholder Perspective and the Relationship Management Project are evidently applicable to 2.0 projects, which are a part of complex projects, but they are also in a completely natural way applicable to the above-defined 3.0 contexts too. In fact, although in portfolios there are evidently both a greater complexity to be faced, and a larger variety of stakeholder relationships to be managed—with certain commonalities among parts of the diverse stakeholders communities, which may be identified by analyzing the eventual correlations among the diverse programs, projects, operations and other activities that are part of the same portfolio, the *value chain of the portfolios* (see Figure 14.1) *has a structure that is almost identical to the project value chain* (in addition, please consider this issue as a cue for further discussions about the actual differences among portfolio, program, and project management). Moreover, almost all main Project Management 3.0 issues that have been someway outlined, including the constant orientation toward business and/or social value, the attention to the coherence between strategies and projects that is represented by

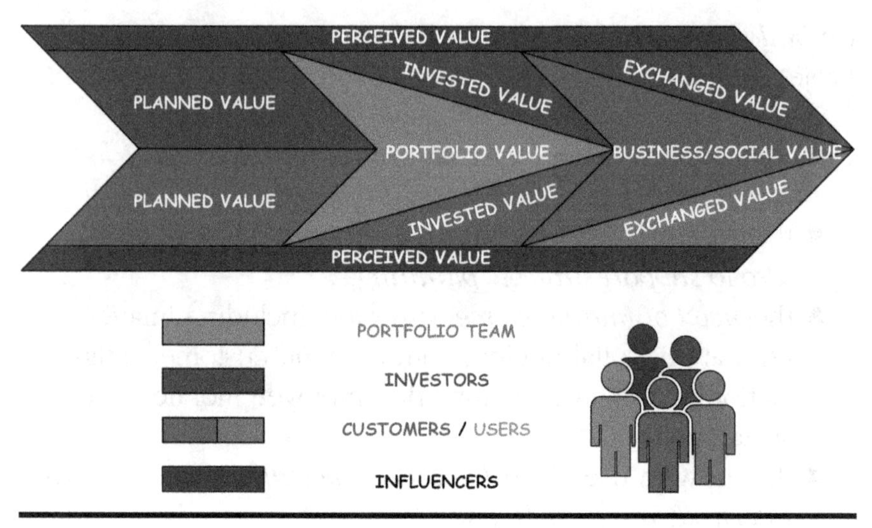

Figure 14.1 A value chain of portfolio investment life cycle.

stakeholder expectations, the use of key performance indicators (KPIs) relevant to both tangible and intangible measures, the capital importance of soft skills, etc., have been extensively developed in previous chapters. In additions, metrics that have been used seem coherent also with recent definitions (Kerzner and Ward, 2019) of Project Management 4.0's Intangible Metrics, and of Project Management 5.0's Strategic Metrics.

Finally, what could be next? Maybe that, in a near and/or medium term future, we will have a Web-oriented *Project Management X.0* (see Figure 14.2), which *will be applicable to both project and portfolio management*, will further benefit from Web's high speed connectivity and distributed software tools, and which:

- *will import from Web 3.0 technologies and developments in terms of Web Data and of Semantic Web, in order both to make available a Global Project Management Knowledge Base of Lessons Learned and to harmonize different languages of very diverse stakeholder communities.* Indeed, the integration and the availability of Project Management Knowledge Bases from global and

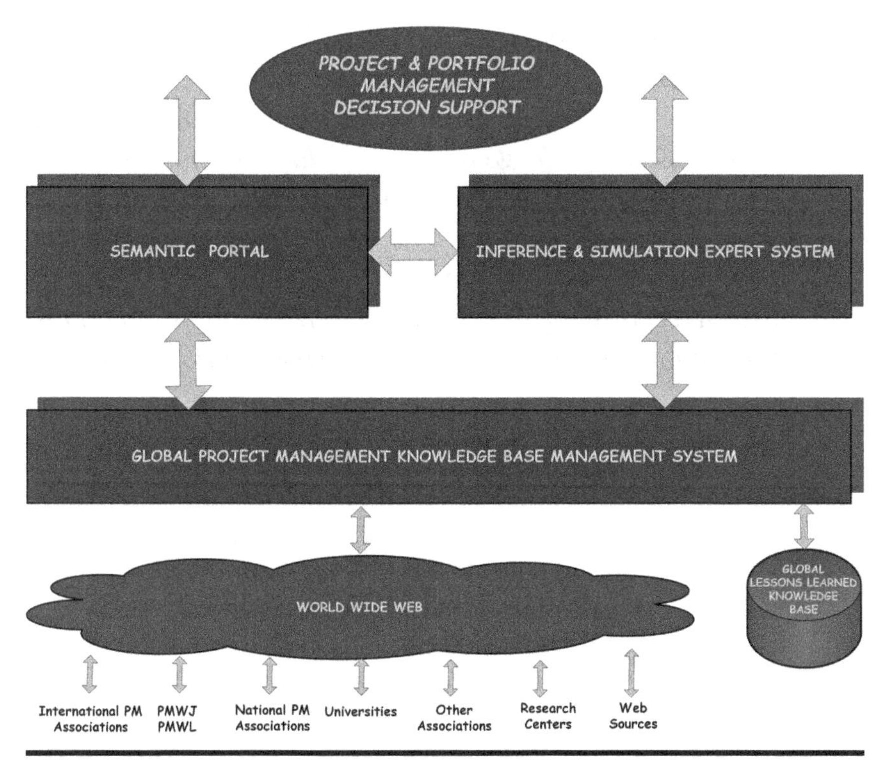

Figure 14.2 Project Management X.0.

specific high-quality sources, including, for instance, PM World Journal and PM World Library, International Project Management Associations as PMI and IPMA, National Project Management Associations as the British APM (Association for Project Management), the Australian AIPM (Australian Institute of Project Management), the Italian ISIPM (Istituto Italiano di Project Management), etc., Universities, Research Centers, Associations that represent diverse industrial/trade/sectors, other web sources, and so forth, *could hugely enhance awareness, knowledge, and cooperation in project management community*; and

- *will import from diverse dimensions of Web 4.0* (Almeida, 2017), including Web of Things, Symbiotic Web, Web Social Computing, and Pervasive/Ubiquitous Computing, those *technologies and developments in terms of media providers, search engines, Internet of Things, Big Data, Artificial Intelligence, Mobile/Cloud Computing, Augmented Reality, devices/sensors, which can allow not an enhanced, effective access to large amount of data from almost everywhere and from a variety of devices, but also the use of powerful inference and simulation expert systems to greatly increase efficacy in decision-making support, so strengthening effectiveness and efficiency in both creation of business/social value and risk management.*

References

Almeida F., November 2017, Concept and Dimensions of Web 4.0, *International Journal of Computers and Technology*, Vol. 16, No. 7.

Archibald R. D., July 2017, On the Stakeholder Perspective, Letter to the Editor, *PM World Journal*, Vol. VI, Issue No. 7, http://pmworldjournal.net/article/on-stakeholder-pespective/, last access on February 2019.

Archibald R. D., February 2018, On the Stakeholder Management Perspective, Letter to the Editor, *PM World Journal*, Vol. VII, Issue No. 2, http://pmworldjournal.net/article/on-stakeholder-management/, last access on February 2019.

Archibald R. D. and Archibald S., 2016, *Leading and Managing Innovation*, 2nd ed., Boca Raton, FL: CRC Press, Taylor & Francis Group, Boca Raton.

Association for Project Management (APM), 2019, *APM Code of Professional Conduct*, Association for Project Management.

Australian Institute of Project Management (AIPM), 2018, *Code of Ethics & Professional Conduct*, Version 5, Australian Institute of Project Management.

Bandler R. and Grinder J., 1982, *ReFraming - Neuro-Linguistic Programming™ and the Transformation of Meaning*, Moab, UT: Real People Press.

Bateson G. and Jackson D. D., 1964, Some Varieties of Pathogenic Organization, in *Disorders of Communication*, edited by David Mc K. Rioch, *Research Publications - Association for Research in Nervous and Mental Disease.*, Vol. 42.

Bourne L., 2015, *Making Projects Work - Effective Stakeholder and Communication Management*, Boca Raton, FL: CRC Press, Taylor & Francis Group.

Cialdini R. B., 2007, *Influence – The Psychology of Persuasion*, New York, NY: Harperbusiness.

Dalcher D., 2018, The leadership imperative and the essence of followership, Advances in Project Management Series, *PM World Journal*, Vol. VII, Issue X.

Doran G. T., 1981, There's a S.M.A.R.T. way to write management's goals and objectives, *Management Review*, No 70.

European Commission – COST Action TU1406 - Quality specifications for roadway bridges, standardization at a European level– WG1, 2016, *Performance Indicators for Roadway Bridges*, Technical Report, July 2016.

European Strategy Forum on Research Infrastructures (ESRI), 2018, *Roadmap 2018 – Strategy Report on Research Infrastructures*, ESRI.

F. Hoffmann-La Roche Ltd, 2017, *Roche in Brief 2016 - Key Performance Indicators of the Roche Group*, F. Hoffmann-La Roche Ltd.

Freeman R. E., 1984, *Strategic Management: A Stakeholder Approach*, Boston, MA: Pitman Series in Business and Public Policy.

Freeman R. E., 1994, The Politics of Stakeholder Theory: Some Future Directions, *Business Ethics Quarterly*, Vol. 4, No. 4.

Freeman R. E., Harrison J. S., and Wicks A. C., 2007, *Managing for Stakeholders: Survival, Reputation and Success*, New Haven, CT: Yale University Press.

Gabassi P. G. (2006), *Psicologia del Lavoro nelle Organizzazioni (Work Psychology in Organizations)*, 7th ed., Milan, Italy: FrancoAngeli.

Glenigan, Constructing Excellence, CITB, and UK Department for Business, Energy, and Industrial Strategy, 2018, *UK Industry Performance Report* - Based on the UK Construction Industry Key Performance Indicators, Glenigan, Constructing Excellence, CITB, and UK Department for Business, Energy, and Industrial Strategy.

Goleman D., Boyatzis R., and McKee A., 2002, *Primal Leadership - Realizing the Power of Emotional Intelligence*, Boston, MA: Harvard Business School Press.

Goleman D., 2011, *Leadership: The Power of Emotional Intelligence - Selected Writings*, Northampton, MA: More Than Sound LLC.

Goleman D., 2012, What is Leadership? http://www.danielgoleman.info/what-is-leadership-3/, Q&A, last access on July 2019.

Hall E. T., 1982, *The Hidden Dimension*, New York, NY: Anchor Books Edition.

International Organization for Standardization (ISO), 2012, *International Standard ISO 21500:2012 Guidance on Project Management - Lignes Directrices sur le Management de Projet*, 1st ed., International Organization for Standardization.

International Project Management Association (IPMA®), 2015, *Individual Competence Baseline for Project, Programme & Portfolio Management*, Version 4.0, International Project Management Association.

International Project Management Association (IPMA), 2015, *IPMA® Code of Ethics and Professional Conduct*, International Project Management Association.

International Telecommunication Union - Focus Group on Smart Sustainable Cities, 2015, *Key Performance Indicators Related to the use of Information and Communication Technology in Smart Sustainable Cities*, International Telecommunication Union.

Istituto Italiano di Project Management (ISIPM), 2017, *Guida ai Temi ed ai Processi di Project Management (Guide to Project Management Themes and Processes)*, edited by Mastrofini E., texts by Introna V., Mastrofini E., Monassi M., Pirozzi M., Trasarti G., Tramontana B., foreword by Archibald R. D., Milan, Italy: FrancoAngeli.

Istituto Italiano di Project Management (ISIPM), 2019, Codice Deontologico (Deontological Code), https://www.isipm.org/chi-siamo/codice-deontologico, last access on July 2019.

Kerzner H., 2015, *Project Management 2.0 - Leveraging Tools, Distributed Collaboration, and Metrics for Project Success*, Hoboken, NJ: Wiley.

Kerzner H., 2017, *Project Management Metrics, KPIs, and Dashboards: A Guide to Measuring and Monitoring Project Performance*, 3rd ed., Hoboken, NJ: Wiley.

Kerzner H., 2019, Project Management 3.0, Free Webinar, https://www.iil.com/free-webinars/Project-Management-3.0/, last access on July 2019.

Kerzner H. and Saladis F. P., 2009, *Value-Driven Project Management*, Hoboken, NJ: Wiley.

Kerzner H. and Ward J. L., 2019, The Future of Project Management, White Paper, https://www.iil.com/resources/future-of-project-management.asp, last access on July 2019.

Marr B., 2012, *Key Performance Indicators: The 75 Measures Every Manager Needs to Know*, Harlow, UK: Financial Times Series.

McQuail D., 2010, *McQuail's Mass Communication Theory*, 6th ed., London, UK: SAGE Publications Ltd.

Mehrabian A., 1971, *Silent Messages*, Belmont, MA: Wadsworth Publishing Company Incorporated.

Mendelow A., 1991, Stakeholder Mapping, *Proceedings of the Second International Conference on Information Systems*, Cambridge, MA.

Mitchell R. K., Agle B. R., and Wood D. J., October 1997, Toward a Theory of Stakeholder Identification and Salience: Defining the Principle of Who and What Really Counts, *The Academy of Management Review*, Vol. 22, No. 4.

Murray-Webster R. and Simon P., November 2006, Making Sense of Stakeholder Mapping, *PM World Today*, Vol. VIII, Issue 11.

Parmenter D., 2015, *Key Performance Indicators: Developing, Implementing, and Using Winning KPIs*, 3rd ed., Hoboken, NJ: Wiley.

Pease A. and Pease B., 2004, *The Definitive Book of Body Language*, Buderim, Australia: Pease International.

Pells D.L., January 2015, Guiding Principles - Commitment to Ethics and Values can Empower Leaders of Teams, Projects, Programs and Organizations, *PM World Journal*, Vol. IV, Issue I.

Pirozzi M., June 2017, The Stakeholder Perspective, Featured Paper, *PM World Journal*, Vol. VI, Issue VI.

Pirozzi M., January 2018, The Stakeholder Management Perspective to increase the Success Rate of Complex Projects, Featured Paper, *PM World Journal*, Vol. VII, Issue I.

PRIME – Platform of Railway Infrastructure Managers in Europe, 2017, *Key Performance Indicators for performance benchmarking*, Version 2.0, PRIME.

Project Management Institute, 1987, *Project Management Body of Knowledge (PMBOK)*, Project Management Institute, republished electronically in 2018 by Max Wideman.

Project Management Institute, 1996, *A Guide to the Project Management Body of Knowledge*, 1st ed., Project Management Institute.

Project Management Institute, 2000, *A Guide to the Project Management Body of Knowledge (PMBOK® Guide)*, 2nd ed., Project Management Institute.

Project Management Institute, 2004, *A Guide to the Project Management Body of Knowledge (PMBOK® Guide)*, 3rd ed., Project Management Institute.

Project Management Institute, 2006, *Code of Ethics and Professional Conduct*, Project Management Institute.
Project Management Institute, 2008, *A Guide to the Project Management Body of Knowledge (PMBOK® Guide)*, 4th ed., Project Management Institute.
Project Management Institute, 2013, *A Guide to the Project Management Body of Knowledge (PMBOK® Guide)*, 5th ed., Project Management Institute.
Project Management Institute, May 2013, *PMI's Pulse of the Profession in Depth Report – The High Cost of Low Performance: The Essential Role of Communications*, Project Management Institute.
Project Management Institute, 2017, *A Guide to the Project Management Body of Knowledge (PMBOK® Guide)*, 6th ed., Project Management Institute.
Project Management Institute, 2018, *PMI's Pulse of the Profession 2018 – 10th Global Project Management Survey – Success in Disruptive Times*, Project Management Institute.
Sampietro M. e Villa T., 2014, *Empowering Project Teams - Using Project Followership to Improve Performance*, Boca Raton, FL: CRC Press, Taylor & Francis Group.
Schramm W., 1955, Information Systems and Mass Communication, *Journalism Quarterly*, Vol. 32, No. 2.
Senge P., 2006, *The Fifth Discipline - The Art & Practice of The Learning Organisation*, 2nd ed., New York, NY: Doubleday.
Snowden D., 2010, An Introduction to Cynefin Framework, https://cognitive-edge.com/videos/cynefin-framework-introduction/, last access on July 2019.
Stretton A., October 2018, A Commentary on Program/Project Stakeholders, Commentary, *PM World Journal*, Vol. VII, Issue X.
Stretton A., December 2018, A Commentary on Strategy Formulation-Related Causes of "Project" Failures in an Organisational Strategic Context, Commentary, *PM World Journal*, Vol. VII, Issue XII.
Transit Cooperative Research Program - sponsored by Federal Transit Administration, 2003, *A Guidebook for Developing a Transit Performance-Measurement System*, Report No. 88, TCRP.
Tuckman B. W., 1965, Developmental sequence in small groups, *Psychological Bulletin*, Vol. 63, No. 6.
Watzlawick P., Beavin Bavelas J., and Jackson D. D., 1967, *Pragmatics of Human Communication*, New York, NY: W. W. Norton & Co.

Index

Note: Locators in *italics* represent figures.

A

B

C

F

G

H

I

Q

R

S

W